JOHN DEERE
Limited-Production & Experimental Tractors

Peter Letourneau

Motorbooks International
Publishers & Wholesalers

First published in 1994 by Motorbooks International Publishers & Wholesalers, PO Box 2, 729 Prospect Avenue, Osceola, WI 54020 USA

Library of Congress Cataloging-in-Publication Data
Letourneau, Peter A.

John Deere limited-production & experimental tractors / Peter A. Letourneau.
p. cm.
Includes index.
ISBN 0-87938-951-6
1. John Deere tractors—History.
2. Prototypes, Engineering—History.
I. Title. II. Title: John Deere limited production and experimental tractors.
TL233.5.L468 1994
629.225—dc20 94-30995

On the front cover: The John Deere Model GPWT.

On the back cover: The Waterloo Boy Model A, a 1919 prototype of the popular and successful Model D. Deere Archives; An AA prototype of the John Deere Model A. Deere Archives; The Model 62, a limited production tractor, followed the Model Y in 1937. *Deere Archives*

Printed and bound in the United States of America

Contents

Acknowledgments

When I approached Motorbooks International with the suggestion that I write a book about experimental and limited production John Deere tractors, I had no idea if I could, in fact, write such a book.

I had no idea as to the amount or extent of information and photographs available concerning such tractors. While I had a vague notion of where to begin, it was nothing more than vague. However, the idea of such a book seemed to be a terrific one, both to myself and everyone at Motorbooks. Before I knew it, I was on my way. Little did I know what lay ahead. In the end, I felt as much detective as writer.

This book is the product of many hours of travel, research, organization, and writing; time spent at the Deere & Company Archives, at the University of California at Davis, and at Worcester Polytechnic Institute in Worcester, Massachusetts.

While the information is sparse concerning some models, I include everything I found of interest. Regrettably, some experimental machines are not included at all, as information was either not available or it escaped me.

While many people contributed information or assisted me in one form or another, my research began and ended at the Deere & Company Archives. Without the cooperation and encouragement of Les Stegh, Deere Archivist, this book could not have been written. I am most grateful for his help.

Foreword

This book is a history of tractor experimentation and on-going development at Deere & Company for the period between 1912 and 1960. It is a story of success and failure; an exciting story of machines that helped make the name "John Deere" synonymous with the term "farm power."

The book is dedicated to the memory of Theo Brown. In 1915, Theophilis Brown designed his first tractor for Deere & Company. In 1916, he was appointed head of the company's experimental department. In 1918, he became Director of Engineering, and in April 1923, was named a director of the company. Brown remained active in the company into the 1950s.

Throughout his career, Brown maintained a daily journal in which he kept a record of his experimental work. He also wrote of the politics, economics, and social events that influenced the routine at Deere & Company. Brown continued his diaries into 1971, the year of his death. Today, the journals are part of the Gordon Library Archives at the Worcester Polytechnic Institute. In August 1992, I spent a week in Worcester, Massachusetts, reading the entries. Beginning with the volume dated 1912, I searched for information of Brown's involvement in experimental tractor design.

The diaries proved to be an excellent source of information and an absolutely remarkable record of Brown's life and work. I am grateful to Lora Brueck, Archivist at the Worcester Polytechnic Institute, for permission to liberally quote from Brown's entries and to reproduce a number of photographs he had preserved, and to Brown's family for permission to reproduce his drawings.

Theo. Brown Has a Diary

Theo. Brown kept a diary, kept it well as it appears;
Seems he kept his diary going through nine rather busy years.
Every evening he thought over what he did that day and said,
Wrote it neatly in his diary and then toddled off to bed.
I have often kept a diary, kept it maybe for a week,
I would then consign my diary to the ash pile or the creek.
Theo. Brown kept his going for nine years without a stop;
When we speak of human wonders, he comes very near the top.

John Deere Magazine
May 1920

Introduction

The history of the American farm tractor dates to the 1840s, when the first portable steam engines were introduced into agriculture. Little more than a steam engine mounted to a farm wagon, the ancestor of today's tractor was primarily used to power the machines of custom threshing operators. Lacking any means of locomotion, the portable steam engine was pulled by horse from farm to farm.

Self-propelled steam engines, or traction engines as they became known, first appeared in the late 1850s. A steam engine which could not only power threshing machines, but also move those machines between farms, appealed to many custom operators. A few manufacturers had more in mind than merely building self-propelled steam power units. Their efforts were directed toward development of a steam plow, a mechanical replacement for the horse and plow. A handful of such machines were built in the 1850s and 1860s but none were successful. Nonetheless, these early efforts to apply mechanical horsepower to plowing were significant steps. Public exhibitions of early steam plows stirred considerable excitement. Encouraged by the response, a number of companies were inspired to develop more practical steam traction units, machines that were suited to field and stationary operations alike.

By the 1890s, steam traction units had improved greatly, and the more reliable they became, the better suited they were to pull heavy draft loads. The decade also saw the introduction of multi-bottom gang plows, specially designed for use with traction engines. These developments greatly enhanced the future of the tractor. Yet, to the overwhelming majority of farmers at the turn of the 20th century, steam traction was too expensive to be employed. However, once less costly gasoline units emerged, and a greater variety of implements were adapted for their use, tractor sales expanded rapidly. By 1920, gasoline powered tractors came to dominate the industry. Use of steam peaked in 1910, and by 1924, the J.I. Case Threshing Machine Company, the world's largest manufacturer of steam tractors, had built its last such unit.

Steam died a quick death once the small gasoline engine became a more practical and efficient supplier of power. Designs of early gasoline tractors closely followed those of steam tractors. In fact, many early gas traction units were built on steam traction chassis and employed similar transmissions and final drives. Consequently, they shared many of the same characteristics, perhaps the least desirable of which was their size.

Early gas tractors tended to be large, cumbersome machines, often referred to as "heavyweights." And heavyweights they were. For example, the Hart-Parr 60-100, built in 1911 and 1912, weighed 26 tons. Its rear drive wheels measured 9ft in diameter. Imagine maneuvering such a monster into a corner of the machine shed! In the mid- to late 1910s, the trend shifted toward smaller tractors, referred to as "lightweights." In general, even these smaller gas tractors were still too expensive to be broadly adopted.

Two developments of the late 1910s and 1920s firmly established the gas tractor as the principal tool of the modern farm. In 1918, Henry Ford introduced the Fordson, the first mass-produced and mass-marketed tractor. Criticized by some as less than ideal, the Fordson's impact was, nevertheless, significant. Its success more than demonstrated the potential of a low-priced, compact tractor. In as little as 10 years, nearly 750,000 Fordson tractors were sold.

The significance of the Fordson notwithstanding, it was the McCormick-Deering Farmall, the first general purpose machine, that made the tractor a viable investment for the average farmer. Because the general purpose, or row-crop tractor, could operate in almost any type of crop, its design became the preferred configuration. Soon after the introduction of the Farmall, almost every tractor manufacturer offered a general purpose model.

Throughout the 1930s and 1940s, manufacturers expanded the range of standard and general purpose models to include larger, as well as smaller and less expensive versions. By 1950, all but the smallest of farmers could afford to own and operate a tractor.

Deere & Company played virtually no role in the early development of either steam or gas traction. Deere's delayed entry into the tractor market is the subject of the first chapter of this book. The second chapter covers the company's decision to build a tractor and examines the experimental work of the period between 1912 and 1918. Successive chapters deal with experimental models built following the purchase of the Waterloo Gas Engine Company.

I have focused on the prominent features of these experimental tractors, rather than dwelling on engineering details. While it was important to me that the book appeal to a broad audience of John Deere enthusiasts, I intended it to be a solid, informative reference work. The topic is fascinating to me, and I hope that the reader enjoys the book as much as I enjoyed researching and writing it.

Chapter 1

Deere's Reluctance to Build a Tractor

The most notable early steam plow was built by Joseph W. Fawkes, in 1858. It generated considerable interest wherever it was demonstrated. John Deere himself took an interest in early steam plows and built a gang plow which Fawkes employed in later demonstrations and, as steam tractors evolved, Deere & Company built plows suited for steam traction.

Deere's First Experimental Tractor

In his book, *John Deere's Company: A History of Deere & Company and Its Times*, Wayne C. Broehl, Jr., revealed evidence of the company's first experimental tractor. In 1858, Deere wrote to a friend that he had ordered a steam plow built in the company's shops. The fact that John Deere experimented with tractors is an exciting thought. Unfortunately, details of this work must be left to our imaginations, as no description, drawings, or other information about this first John Deere experimental tractor exist.

Deere & Company was a leader in the manufacture of farm equipment throughout the period when steam and gas tractors emerged. However, the company took no interest in marketing a farm tractor before 1912 and did not produce a tractor of its own before 1918. Deere's reluctance to build a tractor seems ironic, given its position in today's market. However, it is not difficult to understand the conservative attitude of Deere's management. By way of explanation, in *John Deere General Purpose Tractors 1928-1953: History of the Model GP, A, B, G & H* (Motorbooks International, 1993), I wrote: Consider the following factors:

> First, not everyone in the industry was convinced that the tractor would ever be adopted widely.
>
> Second, prior to 1900, portable steam and gasoline engines were used principally as stationary power units. The primary buyers used their power to drive threshing machines.

For the most part, the tractor industry developed among the builders of threshing equipment—who viewed steam and gasoline traction engines as the logical extension of their existing businesses—and among the engine manufacturers themselves. Deere and Company had no interest in threshing equipment, nor did it build steam or gasoline engines.

A third factor was that prior to the introduction of the Hart-Parr tractor, little practical consideration had been given to the demands placed on an implement when pulled by a steam or gasoline traction engine.

As Fortune magazine wrote, 'the very fact of the tractor gave Deere more to think about than the demand for it.' Implements had to be redesigned and the challenges faced by Deere included the differences in the line of draft [the angle at which an implement was pulled] which, with a horse, sloped upward from the plow bottom to the horse's shoulders; with a tractor, was almost horizontal to the drawbar. Other needs included a means to lift implements over obstructions, while the tractor and implement were in movement; the greater and steadier force exerted by a tractor, which demanded a heavier implement; and the greater cost associated with such revisions.

The state of the art in tractor manufacture was very much in flux, even into the 1920s. The various designs of engines, ignition systems, carburetors, air filters, cooling and exhaust systems, transmissions, final drives, brakes, and so on, confounded even the most experienced mechanical engineers. It was not possible for Deere to simply 'build' a tractor, without considerable investment of time, personnel and money. From the standpoints of logistics and cost, [Deere's] reorganization and expansion of 1910-1912 was a mammoth undertaking that preoccupied Deere's management.

Another factor influencing Deere's conservative approach was that the decentralized structure of the organization left many management responsibilities in the hands of factory and branch managers. Central management was left to the president, an executive committee, and the Board of Directors. A major undertaking like the tractor was not something readily accepted by this diverse group of individuals and interests.

Theo Brown at his desk in 1936. *Deere Archives*

In addition, to execute the reorganization, integration, and harvester expansion programs, Deere had taken on new shareholders (principally, the former owners of the acquired companies and branches) and increased its bank debt. Charles Butterworth [Deere's president], in particular, was concerned that an expenditure on tractor development would be viewed as imprudent by the shareholders and the bankers. Finally, the onset of World War I, diverted interest from tractors, as Deere sales and profits were severely reduced in 1914.

In retrospect, Deere's cautious approach to tractor development is understandable, given all that occurred between 1900 and 1915—at Deere & Company, in the agricultural economy, and within the farm equipment industry.

Chapter 2

The Decision to Build a Tractor

The first record of Deere & Company's commitment to the development of a gas tractor is found in the March 5, 1912, minutes of the Executive Committee. Willard L. Velie, a company director who would later build his own tractor, proposed the following resolution:

"RESOLVED: That in view of the inevitable future use by farmers for diverse purposes of gasoline and kerosene tractors, and especially since the trend is to use them in connection with implements, particularly plows, it seems vital to the interest of this Company that serious cognizance should be taken of the situation, and that through its experimental department, the personnel and talent of which shall be increased, if necessary, a movement to produce a tractor plow should be started at once having in view constantly, that the success of the same would be enhanced if not assured, were it possible to divorce the tractor from the plow and to thus make it available for general purposes."

Velie's resolution gained the support of the Executive Committee. As a result, on July 1, 1912, C. H. Melvin was transferred from his position at the experimental department of the Plow Works to the General Company offices, and given the responsibility of designing an experimental tractor.

Even before Melvin was appointed to his new position, Deere's directors moved to consider alternatives to that of the company designing and building its own tractor. Consideration was given to either acquiring an existing manufacturer or contracting with one to build a tractor under the Deere name. Before examining the Melvin tractor, I think it interesting to review the alternatives considered by Deere's directors.

Gas Traction Company

On March 16, 1912, Floyd R. Todd and C.C. Webber both directors of Deere, met with a Mr. Glover, Director of the Gas Traction Company of Minneapolis, which manufactured the Big Four tractor. Todd reported to the Executive Committee that Glover had been approached by a New York bank which intended to organize a full-line farm equipment company, through the merger of a number of existing manufacturers. The bank sought to include Gas Traction Company as part of the group.

Glover had contacted Webber, General Manager in the Minneapolis branch office of the Deere & Webber Company, to "give us the opportunity to indicate to him as to whether or not we would be interested in his joining the Deere forces, and, if so, upon what basis."

Prior to this meeting, the two companies had already established a relationship of sorts. In 1910, the Gas Traction Company's Big Four 30 won a gold medal for plowing at the famed Winnipeg Agricultural Motor Competition. The Big Four had been matched with a seven-bottom 14in John Deere gang plow. While no closer association between the two companies followed immediately, by 1912, Deere's Atlanta and St. Louis branches did offer the Big Four in their sales catalogs. (It should be noted that the branches had negotiated sales rights, independently from Deere's central management.) The relationship that existed between Gas Traction Company and the St. Louis and Atlanta branches no doubt led Deere's board to seriously consider Glover's advances.

Glover provided a report which indicated that the Gas Traction Company was building fifteen Big Four units a week. It stated that sales had equaled 170 units in 1910; 291 units in 1911; and projected 650 to 750 units in 1912, with estimated net profits of $500,000.

At a meeting on March 21, 1912, the board reviewed Todd's report and went so far as to discuss how many shares the company might offer Glover. It was resolved that George Peek, a member of the Executive Committee and the company's head of sales, should communicate with Webber and outline that:

"(a) In general, we hesitate at jumping hurriedly into a business of this nature, especially as the high profits seem of a temporary character;" yet

"(b) That it is better to acquire a successful going concern, with engines proved in the field, than to develop an engine, factory, and organization, provided such a concern can be acquired on a reasonable basis of common stock."

Further, the resolution directed Webber to confer with Glover and "go as far as necessary to determine if the Gas Traction Company [cared] to go further with the discussion."

Webber did so, and at the board meeting of March 28, reported that "no definitive conclusion had been reached." The board directed that "further investigation of the matter be made, and that Mr. Dain [Joseph Dain, Sr., a director] be requested to give same his attention."

Aultman-Taylor

At the April 9, 1912, meeting of the board, no mention was made of the Gas Traction Company. However, the board did discuss the potential acquisition of Aultman and Taylor, the Mansfield, Ohio manufacturer of steam and gas tractors.

The board concluded that Deere "should seek to obtain reliable information relative to [the] organization and financial condition of the Aultman-Taylor Company, and the merit of the gas tractor made by the company."

While there are references to the merits of Aultman-Taylor tractors in subsequent board minutes, I found no further references which would suggest that Deere, in fact, approached Aultman-Taylor with any form of merger proposal.

Morris Gas Tractor Company

At their meeting of May 7, 1912, the board considered a proposal to buy the Morris Gas Tractor Company of Racine, Wisconsin.

According to C.H. Wendel, author of *Encyclopedia of American Farm Tractors*, (Motorbooks International, 1979), "George Morris had been very successful in his steam engine designs, most of which were purchased by J.I. Case Company. Morris had aspirations of building a gas tractor, and went into business himself."

The directors rejected the idea of the purchase of Morris Tractor Company assets, and approved George Peek's resolution that "the only way we would be interested at present in the Morris Tractor Company, would be in the nature of a selling arrangement, which would not involve any capital on our part."

Seemingly, that was the end of any discussion regarding the Morris tractor.

Diamond Iron Works

Sometime prior to the director's meeting of May 14, 1912, Frank Silloway, who worked under George Peek, met with Mr. Lucke, General Superintendent of the Diamond Iron Works of Minneapolis.

According to Silloway, Lucke had, "come to Moline with the view of interesting us in the tractor which they build, called the American." (Wendel identifies two models built by Diamond Iron Works between 1911 and 1914: a 20-36hp unit and a 40-72hp unit.)

While Silloway's report to Peek and the board expressed no particular interest in the American, it did convey news of interest to the board. In addition to building the American, Diamond Iron Works cut gears for several of the Minnesota-based tractor manufacturers, including Gas Traction Company.

Consequently, Lucke was familiar with their operations, and told Silloway that the Gas Traction Company had sold out to J.I. Case Threshing Machine Company for $1,500,000. (Mr. Lucke's information was not entirely correct. Gas Traction Company had in fact been sold, but to Emerson-Brantingham Implement Company and not to Case.)

Whether or not Deere ever made an offer to purchase Gas Traction Company is unclear. However, the company did act to gain rights to export the Big Four, and Silloway's reaction to the sale leaves little doubt that the company's management was disappointed by the news. He wrote, "As everything has been running along better, and as we did not see how they could afford to turn us down, we have felt a little more secure. The fact, however, that they sold out to Case will make any further introductory work on their engine undesirable on our part."

At the May 14 Director's meeting, it was decided that the company should look elsewhere for a tractor. Peek concluded that "the thing for us to do in order to equip ourselves with tractors for ex-

Above and far right: The Melvin tractor featured an integral three-bottom plow and a power-lift mechanism driven by the engine. *Deere Archives*

port trade, is to have Root & Vandervoot Engineering Company [Moline, Illinois] build us one."

He further stated, "it would be a great advantage to have this tractor built here. . . it would present the advantage that we could build up a trade on this tractor with an assurance that the business would not be terminated by a circumstance similar to that which will undoubtedly terminate our selling arrangement on the Gas Traction Company engine."

The board concluded that Silloway should go to Mr. Vandervoot and "see what he says." Silloway's findings and conclusions, following contact with Vandervoot, were not reported to the directors, as far as I could determine. In fact, discussion of going outside the company to find a tractor does not feature in board minutes again until 1918, when the prospect of purchasing Waterloo Gas Engine Company was first raised.

As mentioned at the beginning of this chapter, on July 1, 1912, C. H. Melvin assumed responsibility for developing a tractor within Deere's experimental department. Perhaps it was the board's frustrations in dealing with Gas Traction Company that ultimately precipitated his appointment. Conjecture aside, let's focus on Melvin's tractor.

The Melvin Tractor

Work on the Melvin tractor was carried out over a two year period, beginning in July 1912. Only one machine was produced at a cost of $6,000.00.

The Melvin tractor closely resembled the Hackney Auto Plow, a lightweight machine

built by the Hackney Manufacturing Company, St. Paul, Minnesota. The Hackney Auto Plow was fitted with a Climax vertical, four-cylinder motor, two-speed transmission, and a patented, power-driven crank system that raised and lowered its three plow bottoms.

Details of the Melvin tractor are sketchy, although photographs clearly display its basic features. The unit was a three-wheeler with power supplied to its two large wheels via pinion gears. A smaller third wheel pivoted to provide the means to steer the tractor. Unfortunately specifications of its engine and transmission are lost.

The tractor was fitted with three plow bottoms, mounted under the tractor frame midway between the front and rear wheels. Like the Hackney tractor, Melvin also incorporated a power-lift mechanism.

The Melvin tractor was designed to be driven in either direction. It was fitted with two seats, one on either side of the centrally located controls. For drawbar work, the smaller wheel ran forward. For plowing operations, the two drive wheels ran forward. During plowing, all three wheels ran on unplowed ground. To facilitate steering and to assist the operator in maintaining the desired cut, the tractor could be fitted with a mounted guide device equipped with one furrow wheel and the ubiquitous directional arrow.

By May 1914, it was decided to suspend further development of the Melvin tractor. There are no records remaining as to its performance. However, almost 40 years after the Melvin tractor was abandoned, Theo Brown wrote, "Joe Dain (Jr.) remembers that the results were disappointing both as to (its) field performance and keeping the tractor from breaking down."

Above and right: Rear and front views of the Melvin tractor. The tractor could be driven in either direction: the larger wheels led when the integral plow was used; the smaller wheel led when pulling. *Deere Archives*

The Sklovsky Motor Plow

In 1912, Max Sklovsky was chief engineer in the general company's experimental department. Prior to Melvin's joining the department, Sklovsky had carried out an evaluation of various tractors on the market. His report was brought to the attention of the Board of Directors by George Mixter on May 23, 1912. (Mixter, great-grandson of John Deere and a graduate of the Sheffield Scientific School at Yale University, directed manufacturing and sat on both the Executive Committee and Board of Directors.)

Sklovsky identified the best tractors in each of three classes: heavy tractor class, Aultman-Taylor; smaller tractor class, Avery; and motor plow class, Hackney.

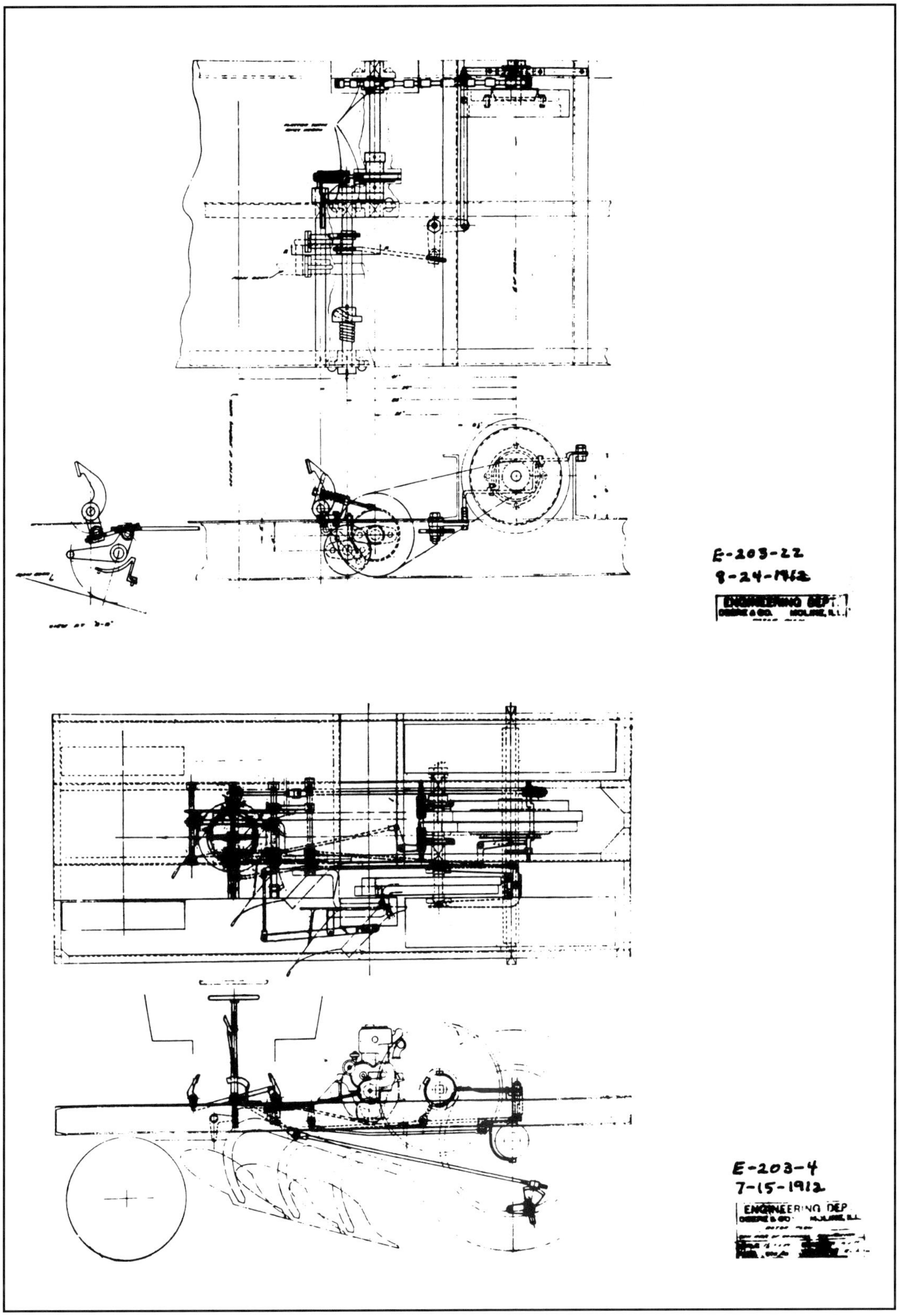

No prototype of the Sklovsky Motor Plow was built. Similar in design to the Hackney Auto Plow, it incorporated a power-lift mechanism driven by the engine. *Deere Archives*

The Melvin tractor pulling a four-bottom plow. *Deere Archives*

The Hackney design impressed Sklovsky, just as it did Melvin. Sklovsky incorporated its general features, including an engine driven power-lift mechanism, into two drawings he completed for a motor plow. The drawings were dated July 15 and August 24, 1912, a matter of weeks after Melvin assumed his new position. While the Deere Board of Directors may have considered Sklovsky's design for further development, Theo Brown wrote that no prototype was built following this particular Sklovsky design.

Chapter 3

Deere's First Limited-Production Tractor

The Dain Tractor

The history of the Dain tractor was well documented and has been preserved at the Deere & Company Archives. It begins in the spring of 1914, by which time Deere had made little progress toward building or acquiring a tractor. On May 27, the Executive Committee called upon Joseph Dain, Sr., to evaluate the situation and to determine if the company could build a suitable three-plow tractor that would sell for about $700.00. At that same meeting, perhaps discouraged by the Melvin tractor, the committee suspended any further work toward tractor development until Dain completed his report.

By June 9, Dain reported that he was working on the design of a tractor that would operate a three-bottom plow. On June 24, the board resolved, "that the preliminary work of designing an efficient small plow tractor be continued under

Dain's first prototype photographed in early 1915. *Deere Archives*

Above: Joseph Dain, Sr. *Deere Archives*

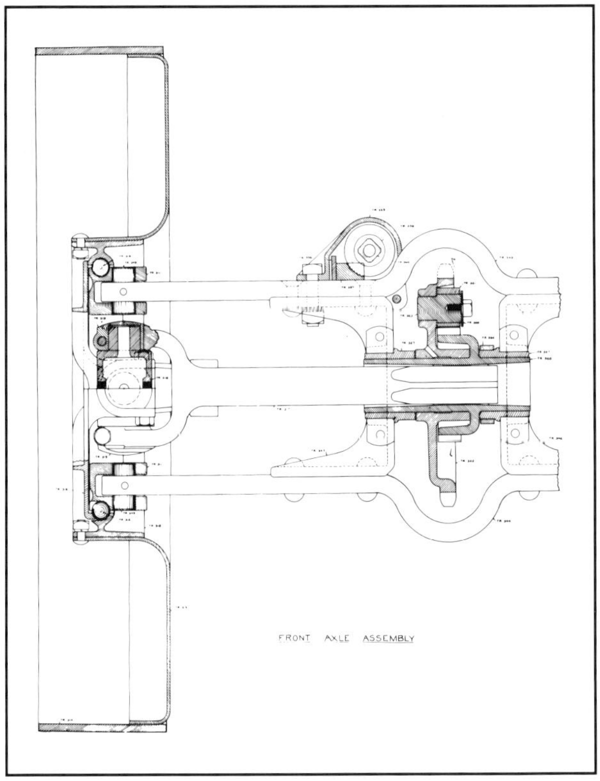

Above: Drawing of the Dain front axle assembly. *Deere*

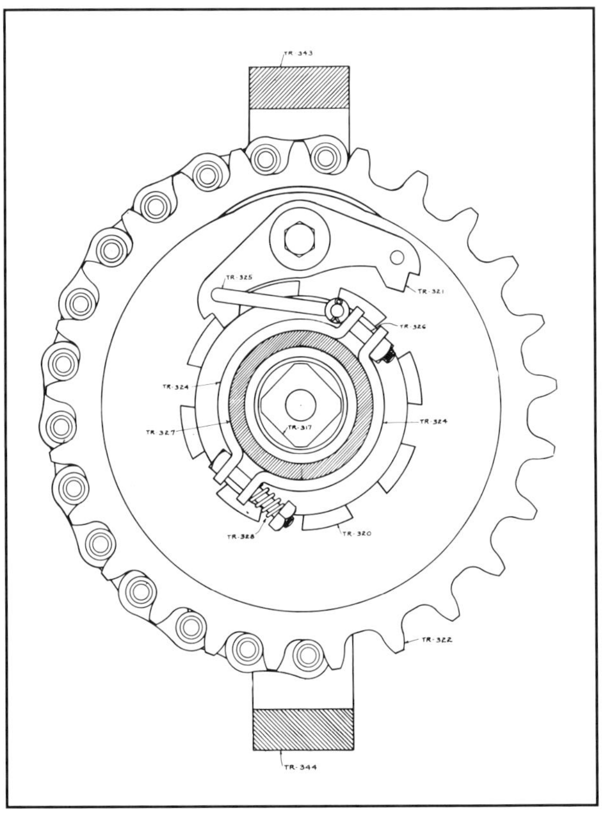

Right: Drawing of the Dain chain-driven wheel and ratchet device that eliminated the need for a differential. *Deere Archives*

the auspices of Mr. Dain and the Experimental Department." On September 8, the board approved Dain's request that $3,000 be allocated to build a prototype.

Dain's First Prototype

By February 1915, Dain's first prototype was completed. However, the board was not yet prepared to consider the Dain tractor for production—not because of its design or any shortfall in performance, but because they were still not certain that Deere should even build a tractor. The board was not reluctant, however, to support continued work on the prototype. It did so, but suggested that the perfected tractor might be sold to another company, rather than be built by Deere.

The valve side of the McVicker-designed four-cylinder Dain engine: (1) water pump; (2) water inlet; (3) governor throttle (disconnected for the photograph); (4) governor throttle arm; (5) governor; (6) air cleaner; (7) oiler drive worm; (oiler drive disconnected); (8) oiler. *Deere Archives*

The minutes from their meeting of February 13 read, "Now, Therefore, Be It Resolved, That Mr. Dain continue his work with the tractor now built until he considers it perfected, and that it be ascertained at that time the amount we have invested, for the purpose of fixing a basis upon which we would sell the same to a separate company for its development and sale. . ."

The board reviewed their position at the meeting of March 9 and lent guarded support to Dain's work, "It was the sentiment of the board that Mr. Dain should continue his work in connection with the tractor, give it a thorough trial in the field, and continue its development; that the present was not the time for Deere & Co. to decide whether it should go into the tractor business or not; that might seem wise later and it might not; that the wise thing for the company to do was to watch the development of this business, and also watch and develop the tractor, in order that the Company might be ready for any emergency."

The Dain prototype was a three-wheel machine, with two wheels at the front and a single wheel at the rear. Equipped with a Waukesha four-cylinder engine, two-speed friction transmission, and chain-driven all-wheel-drive, it weighed about 3,800lbs. Each wheel of the tractor had its own direct power connection. The rear wheel was driven by double chains to sprockets on either side of the wheel. In place of a differential, a ratchet device transferred power from the chain-driven sprocket to the axles that drove the front wheels (see diagram).

Rear view of the Dain: (1) speed shifter rod; (2) speed shifter lever; (3) forward and reverse lever; (4) brake drawbar shifter lever; (5) motor throttle rod. *Deere Archives*

Initial performance tests were conducted indoors. Dain wrote, "We obtained in these tests, by means of artificial traction, a steady drawbar pull of 5,000lbs, on slowest transmission speed." Taken to the field in early 1915, drawbar pull was measured at 3,000lbs under true conditions.

As preparations were under way for full-scale field trials, the directors continued their debate about the merits of building a tractor. Strong pressure in support of the Dain tractor was now being applied from outside the board. By the spring of 1915, a number of field personnel and Deere dealers had reported a disturbing loss of plow sales to competitors who sold tractors. They contended, that if Deere was to continue to hold and expand its market share for plows, as well as other implements and equipment, the company must offer its dealers a tractor.

These dissonant voices were heard by the board and echoed by its own members who supported the demand for a tractor. Minutes from their meeting of May 1, 1915 read, "Mr. Lourie

[Moline Branch Manager] stated that the tractor people are selling plows right on their tractors, and that it is going to interfere with the wheel plow trade."

C.C. Webber called for support for a second Dain prototype and stated that, "The tractor ought to be kept at work to find out the weak places. . ." He further stated that, "with the danger ahead of losing plow trade, as suggested by Mr. Lourie, we ought to be prepared to keep moving." Webber continued by citing opinions of those who had operated the Dain tractor. Two among them stated that they had never seen a tool or an implement that impressed them so favorably.

Left side rear view of the Dain. *Deere Archives*

Lourie then called for at least five tractors to be built and placed to work under various conditions. He also suggested that if the company was reluctant to build tractors because "we had repeatedly told our bankers that we were not in the tractor business," then he felt "it might be feasible to have an independent company organized as a selling company, and to let our factory here produce the tractors and the sales company sell them."

View of the front wheel drive mechanism with the driving plate disconnected and the axle partly withdrawn: (1) Drive plate; (2) Drive axle; (3) U-joint block. *Deere Archives*

Webber suggested that the company go to the bankers and say that "Deere & Co. is going to put this engine on the market; that it is an engine that will pull three plows, and two in difficult plowing; and that it is a good engine, or Deere & Co. would not make it." It was also Webber's opinion that the bankers did not realize that the Dain would be inexpensive enough so that Deere would not have to extend two or three years credit to any farmer who bought one.

Webber and Lourie's appeals apparently moved the rest of the board, as it unanimously adopted the following resolution, "RESOLVED. That without attempting to at this time define our final policy toward the tractor business further than the resolution passed. . . February 13, 1915, but with the view of trying out more thoroughly the Dain tractor, Mr. Mixter be requested to cooperate with Mr. Dain in order to hasten the try-out of the present machine and to determine upon the improvements that should be made in it, with the authority to build from three to six of the revised experimental tractors to be thoroughly tried out this Summer and Fall."

Dain could now continue testing of the first machine, with the knowledge that support for further prototypes had been won. Through the spring and fall of 1915, field tests included plowing and binding operations. Dain reported that the tractor pulled three 14in bottoms in heavy clay sod at a speed of 2.5mph and pulled two binders in heavy oats at 3.0mph.

After more than 300 acres of field work, Dain reported few problems stating, "we have been very well pleased with its performance, considering that, as it is entirely different from any other tractor on the market, we did not have anyone's previous experience to guide us."

The tractor was not without problems, however. Among them Dain cited: a weakness in the drive chains, overcome by putting on heavier chains; failure of the front ratchets, remedied by making them of better material; excessive wear to the gears in the rear wheel; and overheating of the motor, eliminated by changing the radiator and installing a larger size fan.

Dain's Second Tractor

Dain's second prototype was finished in the fall of 1915. Although no record remains of the exact changes it incorporated, a comparison of photographs of the first two machines reveals modifications made to both the steering and operator seating configurations. The basic characteristics, however, remained unchanged: the three-wheel design; the Waukesha four-cylinder engine; the two-speed friction transmission and chain drives. The machine weighed about 4,000lbs.

The second prototype was sent immediately into the field where, operating in wet and heavy soil, it pulled three 14in bottoms six inches deep.

Dain's Third Tractor

A third prototype, completed in December 1915, was practically identical to the second tractor except for its transmission; a positive gear-driven transmission instead of the friction-type. This third tractor was sent to Texas for field trials, and by March 1916, Dain was able to report to the board that it showed "absolutely no weakness in construction." Dain, with the backing of George Mixter, proposed that ten more machines be built at once. The board approved, and by June 1916, six machines were under construction.

These six machines were most likely fitted with the same or a similar Waukesha motor as fitted to the first three prototypes. Reports of poor engine performance were of concern to Dain and Mixter and, in order to overcome problems they had experienced with the Waukesha motor, Deere turned to McVicker Engineering Company, Minneapolis, for a new motor. Designed to Dain and Mixter's specifications, the McVicker engine was fitted to some of the ten final prototypes and was described as "splendid. . . having much more power, but not more than needed, extra strength so it will stand up, and made very accessible."

These final ten prototypes were either loaned or sold to farmers, and were carefully evaluated by Deere personnel. By the end of the 1917 farm season, an extensive report by George Schutz, Minneapolis Branch Manager, cited eight reasons why the Dain tractor would "prove to be a big seller." Included among the reasons were: the three-wheel drive, which allowed the Dain to "go through more difficult conditions without miring down than any other tractor"; maximum traction with minimum weight, its weight being "evenly distributed on all traction wheels, it [packed] the ground less than other makes"; a transmission running in oil, that insured perfect lubrication at all times; "change of speed without clashing of gears"; the four drive chains used to drive the tractor "divide[d] the strain on the sprockets and chains (and made) the chain drive. . . two or three times more durable than gears"; the motor, described as "more durable, being heavier, for its bore and stroke, than any other tractor motor"; and "because of its fool-proof simplicity and accessibility throughout." Schutz concluded his report by calling for the construction of 100 Dain tractors, and on November 19, 1917, the board affirmed the order.

The one hundred Dain tractors were built at the Tenth Street factory in East Moline, between 1918 and 1919. However, full-scale production of the Dain was never ordered. Circumstances had changed. As Theo Brown wrote, Dain had died unexpectedly, before the board had ordered the 100 tractors built; the final price of the tractor was higher than had been expected; but, most significantly, "Deere & Co. had purchased the Waterloo Gasoline Engine Co. and thus had a factory, a

tractor, and an organization that was functioning, and so was in the tractor business at one quick stroke."

Dain Tractor Specifications

Rating: Drawbar, 12hp; belt, 24hp

Speeds: Two forward and two reverse. All speeds direct. No intermediate gear used. High, 2 5/8mph; low, 2mph. Reverse speeds same as forward.

Weight: 4,600lbs

Length: Wheel base, 6ft 2 3/4in; overall, 12ft 6in

Height: 4ft 9in

Width: 6ft 4in

Wheels: All traction. Rear, 40in diameter, 20in face; front, 36in diameter, 8in face.

Steering gear: Worm and sector.

Transmission: Exclusive design. No gears to shift in changing speeds. Change can be made from low to high or vice versa while under full load without stopping the machine.

Bore and Stroke: 4.5x5.0in

Crankshaft: 3 bearings, 2 1/2in in diameter. Drop forged from open hearth steel and double heat treated.

Bearings: Connecting rod, 2-1/2in diameter by 2 7/8in long. Front main, 2 1/2in diameter by 3 7/8in long. Center main, 2 1/2in diameter by 3in long. Rear main, 2 1/2in diameter by 3in long. All bearings bronze back with special babbitt metal lining.

Lubrication: Force feed from main oiler, supplemented by a splash system.

Governor: John Deere design, built in.

Cooling: Water circulated by positively driven centrifugal pump of extra large capacity.

Fan: 19in in diameter. Positive gear drive.

Carburetor: 1 1/4in Model M-2 Stromberg.

Air Cleaner: Bennett.

Ignition: KW high tension magneto with starter coupling.

Belt Pulley: Steel, split hub, 30in diameter with 8in face. Belt speed, 2,190ft per minute.

Above and far left: Dain's second prototype was completed in the fall of 1915. *Deere Archives*

Prototype No. 3 undergoing tests in Texas. *Deere Archives*

Above: Dain tractors lined up in the streets of Huron, South Dakota. *Deere Archives*

The Dain tractor drawing a plow, circa 1919. *Deere Archives*

Chapter 4

Two Plow Tractors and Various Motorized Cultivators

Sklovsky's A2, B2, and D2 Tractors

At the same time that Joe Dain worked to develop his three-plow tractor, Max Sklovsky, at the direction of George Mixter, undertook the design of a two-plow tractor.

His first effort was the Model A-2, an all-wheel-drive, three-wheel tractor, employing a Northway four-cylinder engine. The tractor had no differential. Its front axle pivoted in the center like a wagon axle. The most distinctive feature of the A-2 was its one-piece cast-iron construction. The frame, incorporating the engine pan, was cast as a single unit and, according to Theo Brown, was the industry's first such design. (While Deere never employed this form of construction in a production tractor, J.I. Case used it in its crossmotor models 9-18B, 10-18, 12-20, 15-27, and 18-32.)

While the tractor generally performed well, its steering was problematic. Theo Brown wrote of its first demonstration for Deere executives on November 20, 1915, "the exhibition was not particularly good as the steering was hard." He later described it as "almost impossible. . . due to having the power-driven front wheels mounted on a wagon-type axle and with no differential."

The A-2 was built at the Marseilles plant in East Moline in the fall of 1915, at which time Brown was its superintendent. George Mixter put Brown to work on the A-2. He attempted blocking the wheels, so that they were solid on the axle, but this was not the solution. In the end, Mixter concluded that an automobile-style axle and steering were necessary.

A revised tractor, designated the model B-2, was built. Its performance was "quite satisfactory," according to Brown. However, the B-2 was not deemed commercially practical. The high cost of its Northway engine pushed estimated production costs above a level acceptable to Deere.

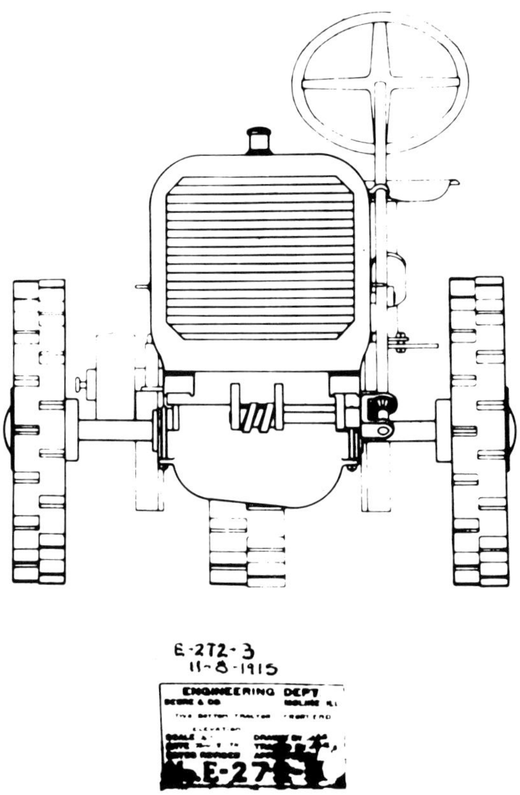

Above: The A-2, like the Dain tractor, featured all-wheel drive. Its entire frame, including engine pan, was a single casting. *Deere Archives*

Opposite page: Only drawings of the two-plow Sklovsky A-2 tractor remain. A prototype was built in 1915. Its clutch, steering, and front axle were problematic. *Deere Archives*

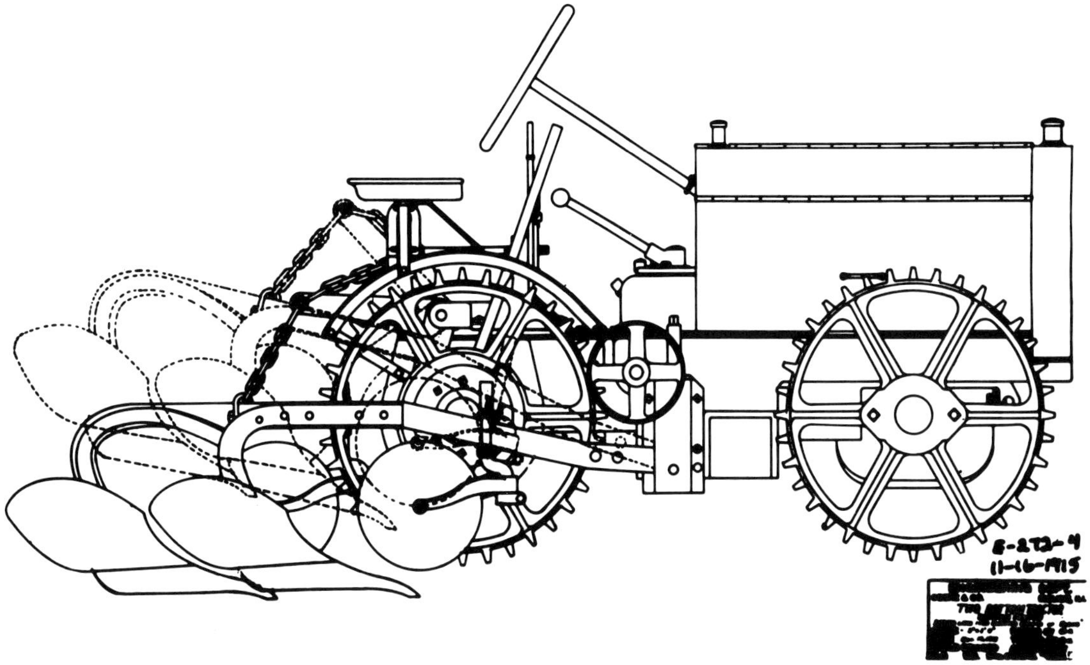
E-272-4
11-16-1915

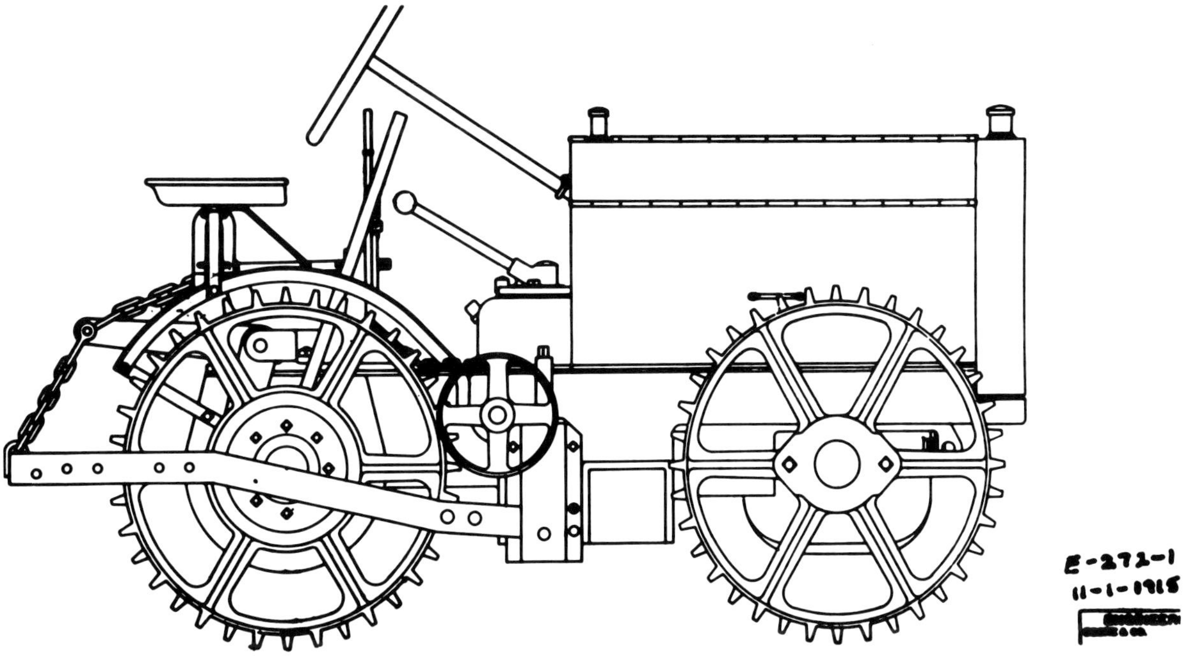
E-272-1
11-1-1915

Max Sklovsky seated on the B-2 prototype. The B-2 was considered for production as a companion to the larger three-plow Dain tractor. *Deere Archives*

Sklovsky's design did not die immediately, however. Rather, the tractor was redesigned with a single-cylinder, hopper-cooled engine. This engine had the advantages of low cost, easy accessibility, and the fact it could burn kerosene rather than gasoline, as was necessary for the Northway four-cylinder. The tractor, designated the D-2, did not get beyond the drawing board stage, however, as the demands placed on the company by its war work for the U.S. and allied armies of World War I took priority.

Theo Brown's Tractivator

In September 1915, Theo Brown was offered the job as head of Deere's experimental department. The appointment was confirmed by the Executive Committee in November, and Brown moved from the Marseilles Plant into his new office in late February 1916.

Brown's first major project was the design of a one-row motor cultivator. As background to the project's evolution, Brown later wrote, "The [industry's] first attempts to design a tractor for cultivating row crops followed the pattern established for horse-drawn cultivators, except that a steering wheel replaced the harness reins. On both horse-drawn cultivators and early motor cultivators, the operator steered the outfit with his hands and manipulated the cultivating rigs with his feet.

"A troublesome difference between horses and tractors immediately became obvious. On horse-drawn cultivators, the operator's double job was easy, for horses soon learned to follow the rows, permitting the operator to give his full attention to dodging the rigs. But if tractors didn't grab an occasional mouthful of corn leaves, neither did they learn to follow the corn rows. So cultivating, which had been a comparatively easy job with good horses, suddenly became tough, exacting work with the early motor cultivators, for the operator had to perform two unrelated functions simultaneously—looking ahead and steering the tractor, and looking down and manipulating the rigs. . . this type of motor cultivator proved to be impractical, because a tractor operator could not successfully coordinate two unrelated functions."

By the end of March 1916, Brown had sketched the design of a motor cultivator on which the front pivotal wheels worked both the rigs and steered the tractor, thus eliminating the operator's need to coordinate the job of steering and manipulating the rigs. Work on a prototype followed and by June 1 it was in the field. Brown wrote that "the first tryout was a success."

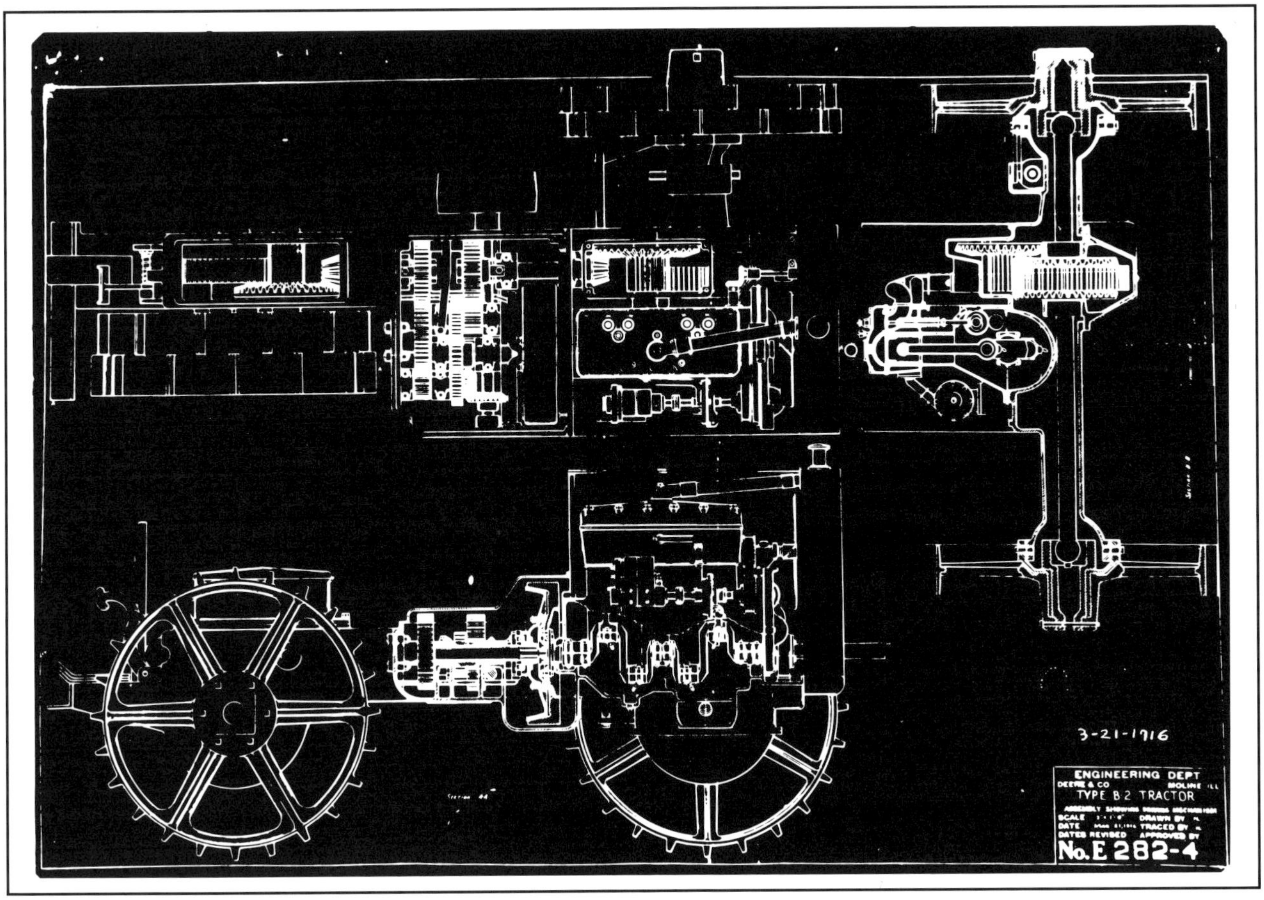

A blueprint of the Model B-2 shows the direct gearing of its front and rear drive mechanisms. No differential was employed. *Deere Archives*

Max Sklovsky at the wheel of the B-2 prototype. Its four-cylinder Northway engine was considered impractical. *Deere Archives*

The prototype was assembled from a K.C. cultivator, a frame supported by manure spreader wheels, spreader drive sprockets and chains, and a 7 1/2hp, single-cylinder New Way air-cooled engine. The unit weighed 700lbs, less the engine. By July, it was decided to fully develop the motor cultivator for production. Work continued on the design, with minor improvements made to its performance. A new one-speed transmission and gear drive were developed, and French and Hecht designed new wheels. The new prototype was tested in the field throughout the 1916 season.

Field trials demonstrated the need for a more powerful engine. Subsequently, several engines were tested. The first, a 4 1/4x6in, single-cylinder engine from Associated Manufacturers Co. was, according to Brown, "entirely too light and would not pull at all "Associated's engineer proposed to build a larger 5 1/4x8in, single-cylinder engine. This larger engine, a New Way horizontal engine, and a Knight engine were also tested. In the end, Deere called upon McVicker Engineering, who designed a two-cylinder, hopper-cooled engine that was built by Associated Manufacturers. Kingston was contacted to supply carburetors and Dixie

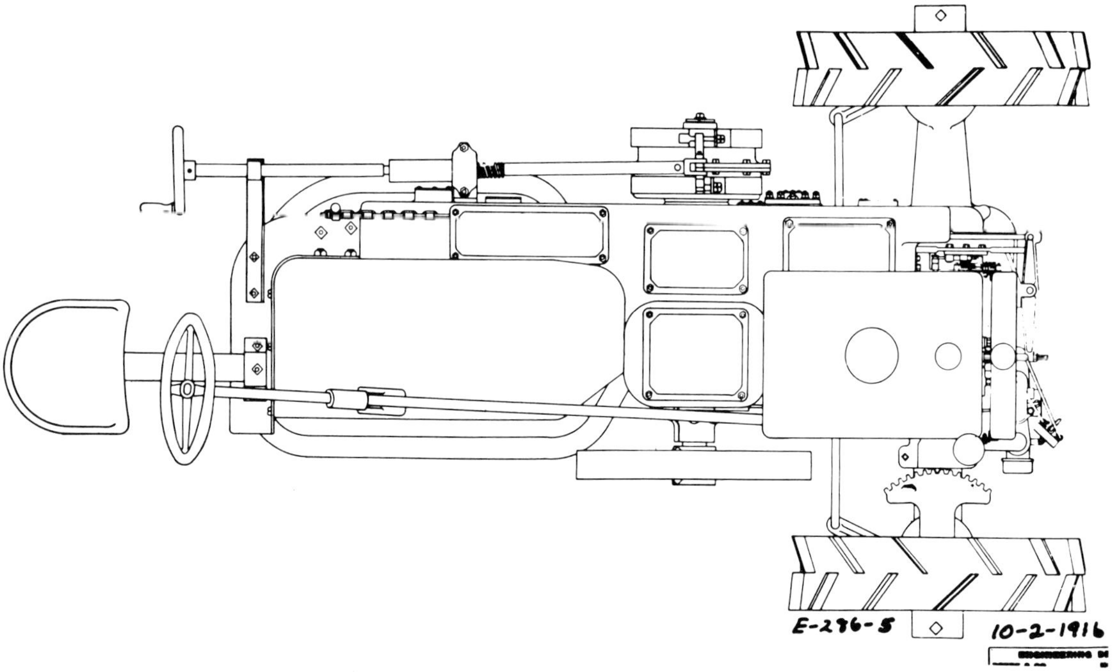

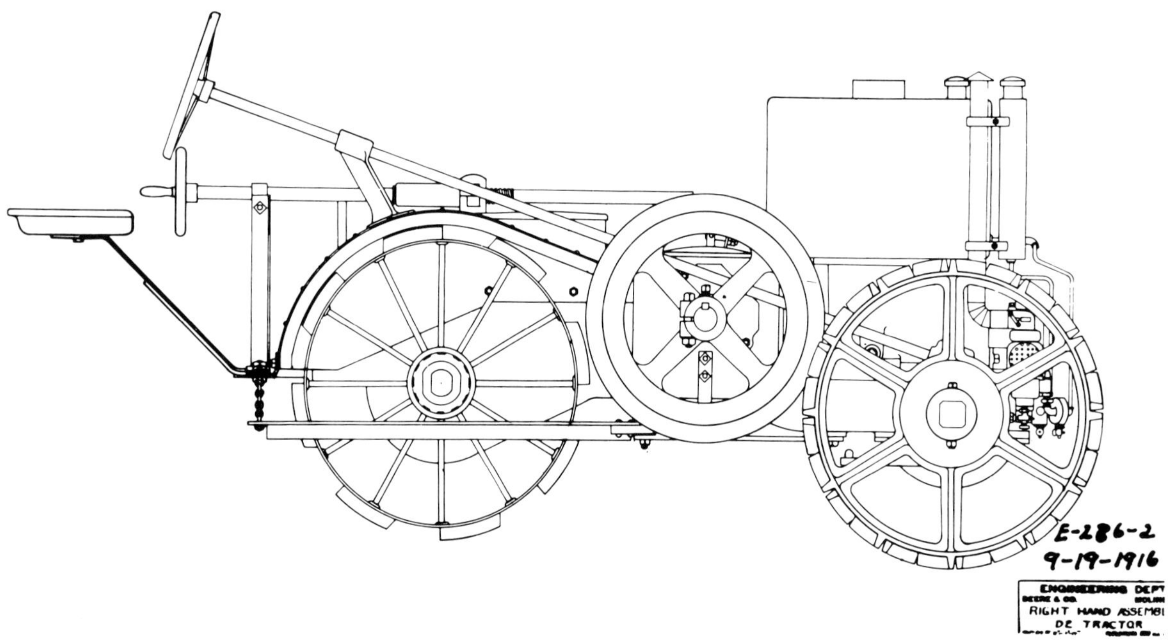

Above and opposite: Side, top, and front views of the Sklovsky D-2. A single-cylinder, all-wheel drive tractor, it made it no further than the drawing board. All work stopped on a two-plow tractor during World War I. *Deere Archives*

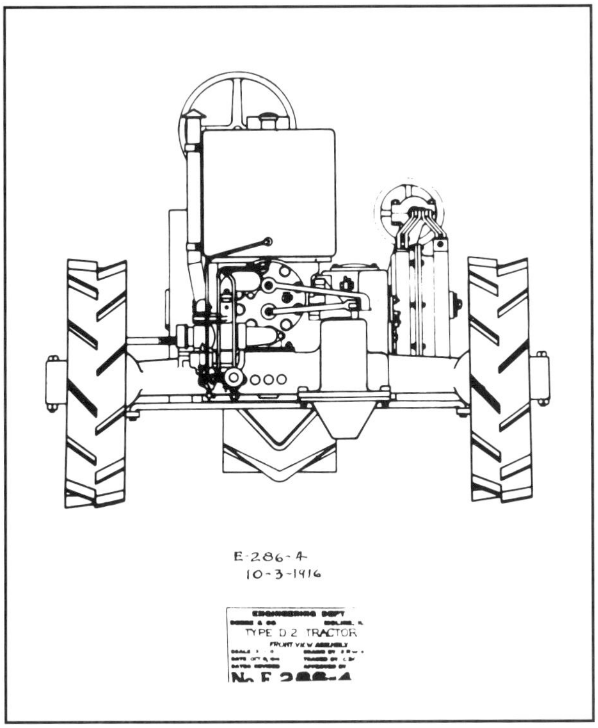

supplied magnetos. The new engine weighed 510lbs and gave the unit an overall weight of about 1,200lbs.

As the new engine was being developed, Brown worked with Deere engineers to test other implements for use with the motor cultivator. The prototype was modified to run in the opposite direction for plowing. A special hitch was designed and binding, disking, and mowing operations were attempted with success. A two-row cultivating attachment was designed, but was not successful.

On January 16, 1917, the Board of Directors approved the purchase of materials to build 50 of Brown's tractor cultivators, 25 machines to be built immediately and another 25 machines to be built after further trials. By April, the machine had gained the designation "Tractivator"; Brown was replaced by L.R. Clausen as head of the project; and all 25 units were assigned to various branches for sale.

Once the Tractivators were delivered to customers, Deere field personnel were assigned to each machine to report daily on their performance. Brown wrote, "The reports clearly indicate that the Tractivator as a cultivating outfit performed satisfactorily both as to quality of work and ease of operation. Thus, the basic idea of shifting the cultivator rigs in dodging corn and steering the tractor with one and the same operation was proved to be sound.

"However, the reports revealed that a man with the Tractivator did not accomplish any more work in a day than he could with a pair of horses, and for that reason he could not afford to buy the Tractivator."

Because the Tractivator could do no more work than a team of horses, the 25 units built were returned to the factory. Deere dropped the notion of building a one-row cultivator, but immediately took up development of a two-row machine.

Theo Brown wrote, "By the middle of June 1917, reports from the field on the 25 one-row Tractivators had pretty well established three important facts:

1. One-row motor cultivators probably would not be economically successful, for farmers were finding that they could cultivate just as much ground in a day with a team of horses.
2. Row crops could, however, be successfully cultivated with mechanical power.
3. Therefore the logical step would be to develop a two-row cultivator."

Silver's Cultivators—Models No. 1 and 2

Walter Silver developed the company's first two-row motor cultivator in June 1917, using an Avery motor with friction transmission. Unlike the four-wheel Tractivator with a rear-mounted engine that supplied power to the rear wheels, Silver's machine was a three-wheeler with a front-mounted engine that powered the front wheels and a small rear wheel that steered the unit. The operator shifted the cultivator rigs with his feet and separately steered the unit with a steering wheel.

Brown wrote, "as with previous motor cultivators, it soon was evident that requiring the operator to perform two unrelated functions—steer the tractor and manipulate the rigs—was not satisfactory."

Silver built a second prototype with steerable front wheels that solved the problem. The small rear wheel could also be steered to help in turning around at the end of a row.

Models No. 3 and 4

In September 1917, the Board of Directors approved continued development of Silver's motor cultivator. During the winter it was completely redesigned. A third model prototype was built by mid-April 1918 and sent to the field for trials. By June, Silver made further changes to the design, and five examples of his Model 4 were built and sent to farmers for further testing.

Above and opposite page: Brown's first single-row motor cultivator of June 1916 was powered by a New Way, 5 1/4x6in, single cylinder air-cooled engine. Its wheels, drive sprockets, and chains were borrowed from a Deere manure spreader. *Deere Archives*

Above and right: The first of twenty-five production single-row Tractivators built in 1917. It was powered by a McVicker designed, 3 7/8x5 3/4in, two-cylinder, hopper-cooled engine. *Deere Archives*

Although the Tractivator performed well, it could do no more work than a team of horses. *Deere Archives*

Among the farmers who tested Silver's Model No. 4 was J.B. O'Donnell of Sheldon, Iowa. O'Donnell, who also tested Model No. 2, was employed by Deere as a consultant. He subjected the motor cultivator to extensive testing, using it to disk, plant, cultivate, mow, pull a binder and a 20-foot drag harrow. His report to Deere was very positive, recommending that a larger motor of at least 6hp be used, and that proper hitches be developed for different implements.

Effectively, O'Donnell recommended that the motor cultivator be redesigned as a "universal tractor." He wrote, "There are more farmers in the market today for this type of tractor than any other machine of its kind. I have talked to hundreds of farmers in the past two years. . . the farmer is interested in and looking for a more universal type of tractor than he has been able to get so far."

O'Donnell concluded, "In designing this machine, it should be remembered that it is not only a motor cultivator, but an all-purpose tractor, and that the farmer will have work for it to do practically every month in the year."

Despite O'Donnell's glowing reports, little more was done regarding the Model 4 motor cultivator, and little information is available to explain why. The March 1918 acquisition of Waterloo Gas Engine Company was most likely the reason for the company's sudden loss of interest. Deere now had both a tractor and experienced management to build it. Much of the focus on tractor development shifted to Waterloo, while many of those who remained in Moline returned to their work developing implements.

In September 1918, the Board of Directors recommended that the five Model 4 prototypes be tested through another season, and in December 1919, motor cultivator development work was transferred to the Plow Works. With the exception of a single-row cultivator built in 1921, no further consideration was given to building a motorized cultivator.

The 1921 Single-Row Motor Cultivator

In early 1921, a prototype one-row cultivator was built following the lines of Silver's first two-row cultivator of 1917. According to Theo Brown, "the cultivator rigs were shifted laterally by foot pedals, and the tractor was steered by guiding the two rear caster wheels. Thus, there were two distinctly separate functions for the operator to perform."

The cultivator prototype received a cool reception from Deere's Board of Directors. As Brown noted, the two distinct operations required of the operator was "a step backwards" in design. In October, the board acted to stop all further development on motor cultivators.

According to Brown, there were several reasons for such action: the depression of 1921 caused the board to scrutinize every expenditure closely, with only the most urgent given any consideration; the re-design of the Waterloo Boy tractor, the Model D prototype, was given priority over all other experimental work; the two-row cultivator was not adaptable for plowing.

The concept that a cultivating tractor should also adequately handle a plow was, of course, the very notion that led to the development of the general purpose tractor. Farmall, the industry's first general purpose tractor, had evolved from International Harvester's less-than-successful efforts at building motorized cultivators. It is ironic that the Deere's decision to drop further work on motor cultivators coincided with International Harvester's decision to build the first 50 Farmall prototypes.

The Mystery Tractor

While detailed records of experimental work were sometimes preserved, in many cases little but photographs remain today. One such example is what I like to call the Mystery Tractor.

Silver Motor Cultivator No. 3 equipped with a Model 999 planter. April 1918. *Deere Archives*

Closely resembling the Dain tractor, the Mystery Tractor may indeed have been constructed by Dain. However, records in the Deere Archives suggest that it may have been the work of Max Sklovsky. It is interesting to note that in February 1916, Theo Brown was given the responsibility of getting Dain and Sklovsky together to develop a tractor incorporating features from their individual designs. Perhaps the Mystery Tractor is a result of that effort.

Above and opposite page: Silver Model No. 4 incorporated changes to the power driven cultivator wheels. Deere Archives

Left and above: Front and side views of the 1921 One-Row cultivator. Note the foot pedals used to shift the cultivator rigs from side to side, and the two rear caster wheels. June 10, 1921. *Deere Archives*

Above and following pages: Dain's work or not? On the one hand, there are similarities in the frame, front wheel hubs, wheel spoke design, and an apparent friction transmission, but on the other hand, there are two rather than one rear wheel, as well as a different operator position. No details accompanied these photos. *Deere Archives*

Similar in appearance to the second Dain prototype, this may have been a Dain design, or possibly the work of Max Sklovsky. *Deere Archives*

Chapter 5

Model D: A Successor to the Waterloo Boy

Above and left: Two views of the Waterloo Boy Model A, a 1919 prototype of the Model D. Its wheelbase was longer than that of the design illustrated in the drawings of December 1920. *Deere Archives*

Developments that led to the production of the Model D began prior to Deere's purchase of Waterloo Gas Engine Company. Correspondence in the Deere Archives reveals that engineers at Waterloo Gas Engine Company made preliminary sketches for a fully enclosed-gear tractor as early as the fall of 1915. However, no engineering drawings were executed before 1916, and the first such prototype, dubbed the "bathtub" tractor because of the shape of its main case, was not built until the spring of 1917.

The Bathtub Tractor

I found no photographs or drawings of the 1917 prototype, however, a fairly detailed description of the tractor remains in the Deere Archives, excerpts of which follow:

"Engine was primarily the same as the Model 'N' tractor engine, with cylinder head to the front end of the tractor and with exhaust let out to the front underneath radiator. . . Engine speed was 750rpm, the same as that of (Waterloo Boy) Models 'R' and 'N'.

"Channel iron was used to make the front end of the tractor. It was bolted to the bathtub case and on it was mounted the water pump, fuel tank, and radiator. . .

"The front axle was mounted underneath and to the channel iron front end support, and was a modification of Model 'N' front axle.

"Power was taken from the engine by means of contracting band clutch and was delivered to a sliding gear shaft by what is now termed 'First Reduction Gears'. . . This model was provided with two forward speeds and one reverse. . . Final Drive was by means of large, wide gear integral with rear axle shaft."

This first prototype was not satisfactory, as there were problems with the all-gear final drive. The reports stated, "Because of having to leave a large opening in the top of the bathtub case to install these final drive gears, the case was very

The Model A, known as the Bathtub Tractor because of the shape of its main case. *Deere Archives*

weak at this point and deflected considerably—so much that the teeth on the differential compensating gear and final drive gear could not be made to operate at full contact."

A second prototype was built in the summer of 1917. It was similar in all respects to the first, except the final drive of a Model 'N' was employed. The records indicate that this tractor was no more successful than the first "because the bull pinion and final drive gears were exposed and subjected to a lot of dirt. . . they wore very rapidly." The problem of final drive wear was eventually overcome by using sprockets and roller chains running in a bath of oil, which gave a flexibility required to overcome the deflections of the casing.

The "Pre-D" 1919-1922

Beginning in 1919, the first of three series of prototypes that incorporated roller chain final drives was built. The first series of six tractors, numbered 101 to 106, were built and tested extensively around Moline and Waterloo. Between September and December 1921, a second series of six prototypes numbered 201 to 206 were built and tested in Moline, Waterloo, and Jamestown, North Dakota. The third series of prototypes followed in the summer of 1922. There is some confusion over the number of tractors built in this third series. The report on file at the Deere Archives indicates two tractors, numbered 301 and 302; Don Macmillan and Russell Jones, in their work, *John Deere Tractors and Equipment, Volume One 1837-1959*, (American Society of Agricultural Engineers, 1988), state twelve; the Two-Cylinder Club publication, *Two-Cylinder* claims eleven, numbered 301 to 311. There is also some question as to the model designations used by the company to differentiate between the three series of pre-D tractors. The first series was referred to as the Model A, which may well have been the only designation used. Some claim that the second series was labeled the Model B, the third series the Model C, and the production tractor the Model D.

Principal characteristics of the various prototypes are noted below in excerpts from the report of the Model D's development on file in Moline, into which I have interjected relevant comments from Theo Brown's journal:

"The engine speed was increased from 750 to 800.

"The cylinder block was attached to the front end of the bathtub case and the crankcase became an integral part of the bathtub.

"The front axle and the radiator were carried on a frame which extended out from the bathtub case. This was so designed because at the time Fordson carried their axle on a bracket hooked to the engine and had a patent on it.

"There was a lot of work done to shorten the tractor, so that it could be shipped crosswise in boxcar, which was accomplished. . . "

Theo Brown's journal entry of January 13, 1921, read, "In afternoon went out to the Marseilles Works with Clausen to go over the new tractor design. It is beginning to take a more pleasing appear-

The 1919 Bathtub Tractor pulling a three-bottom plow. *Deere Archives*

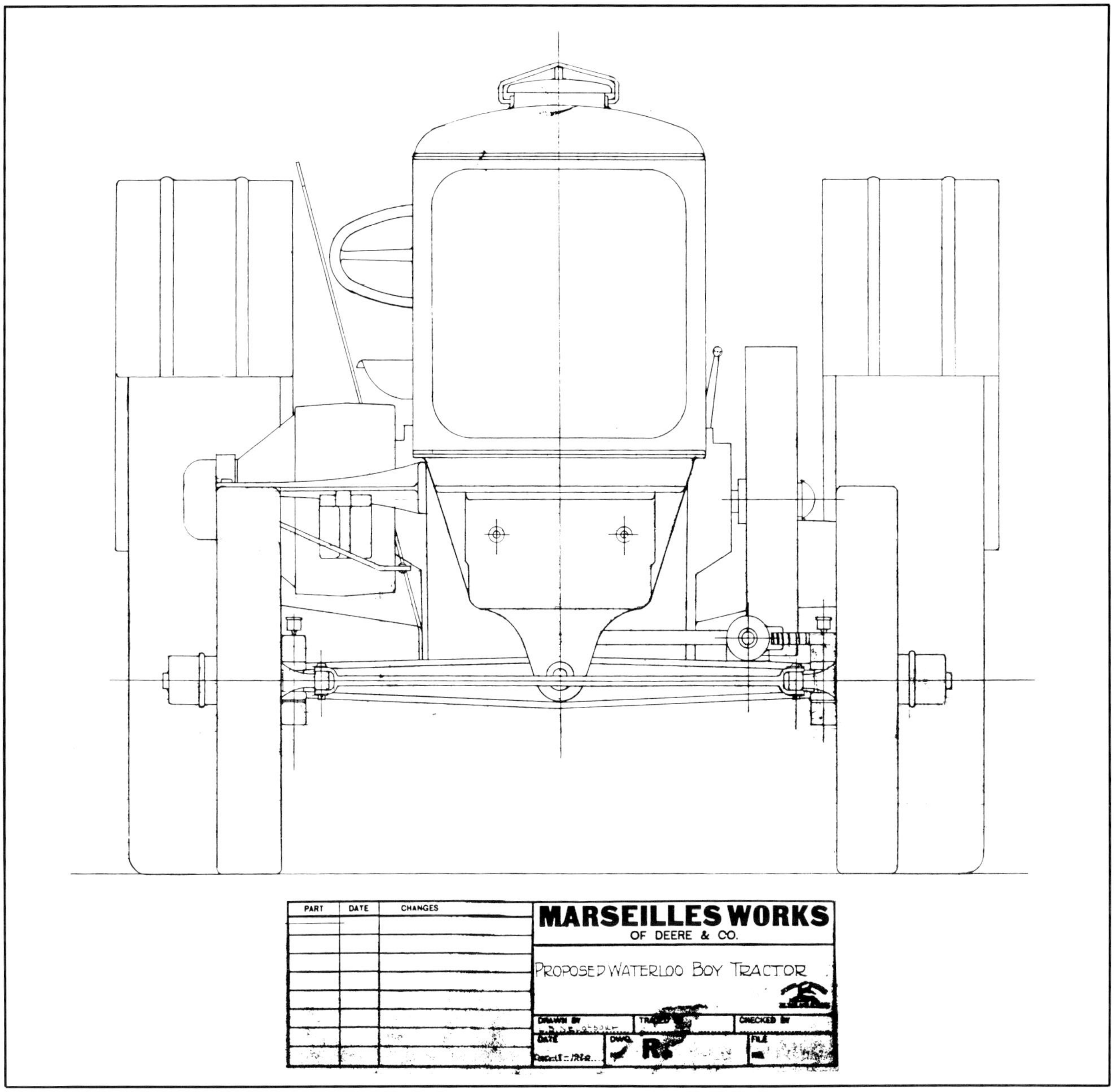

Above and next page: Front and side elevations of the Model D prototype, as of December 1920. *Deere Archives*

ance now. We have been able to lower the height and take 12in off the length of the wheelbase."

"The extensive program carried on to reduce the weight by keeping all castings to the minimum thickness accomplished its purpose, in that the machine weighed under 4,000lbs, the goal at which they were shooting."

Theo Brown's journal entry of June 16, 1921, recorded the weight at 3,950lbs. Three weeks later he wrote, "In afternoon went out to see the new tractor and an old Waterloo Boy plowing in sand. The old Waterloo Boy being heavier had better traction than the new tractor.

"Along with the light weight, it was necessary to develop a means of getting good wheel traction with a limited amount of weight—the angle iron grousers, which had been used up to then, would not penetrate to a satisfactory depth to hold the engine—slippage due to shearing off the topsoil was very high—a spade lug 4in high was devel-

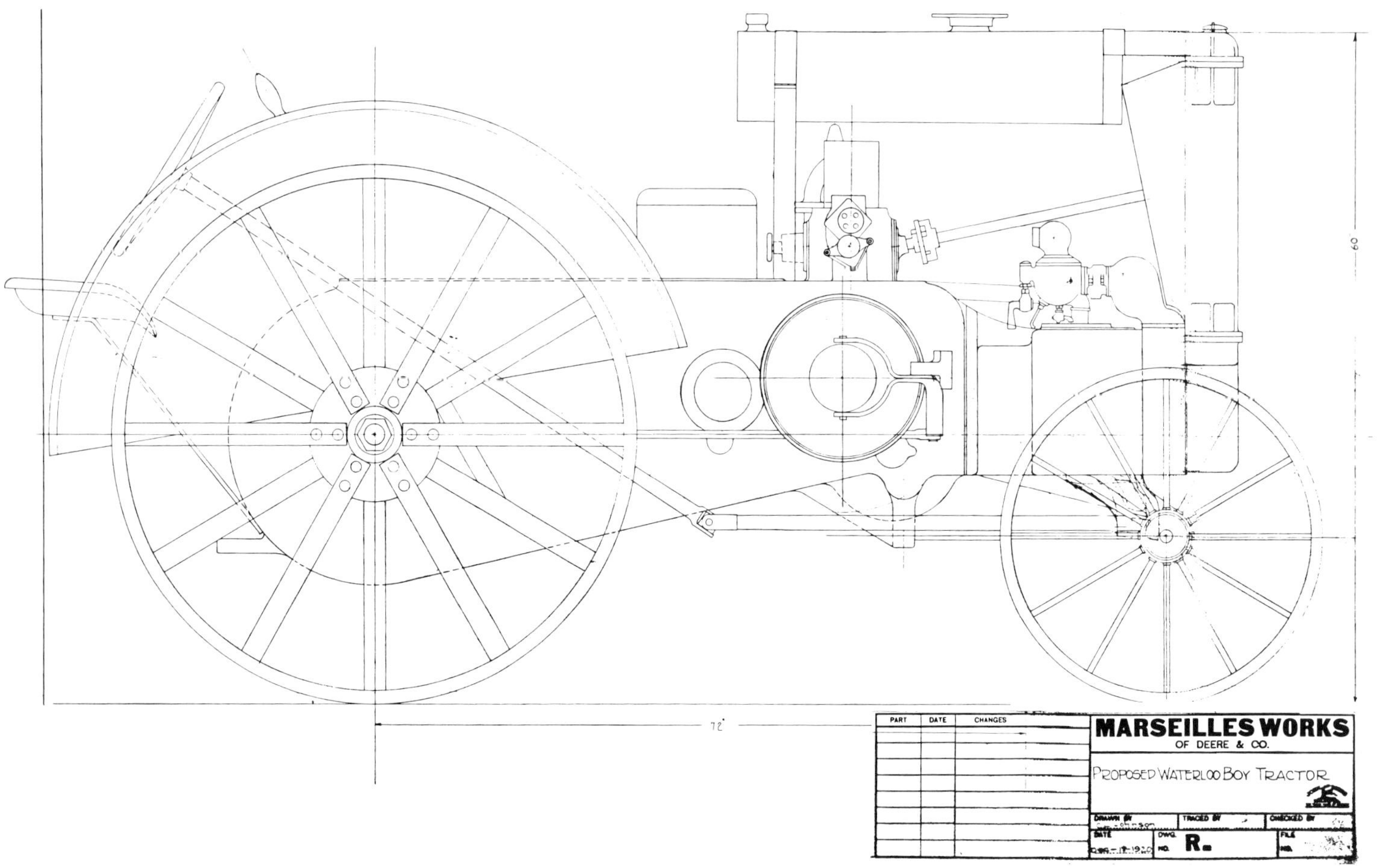

oped—the reduced edge contact with the ground allowed lugs to penetrate to a depth sufficient to hold engine power in second gear; with, but very little slippage in good loam soil-slippage was not objectionable in low gear in most soil conditions when plowing. Longer spade lugs—5in and 6in—were later added for loose soil conditions.

"One of the basic changes that took place in this period was the changes of the engine intake and exhaust system to burn various kinds of kerosene. . . The sum and substance of it was that the intake and exhaust systems were combined and brought to the top of the cylinder head into one manifold, in order to get enough heat on the intake to properly handle various kinds of kerosene. We didn't know anything about distillate in those days. . . This system worked out very good except that it became necessary, as we put more heat on the intake, to feed the engine some water to keep down preignition. . .

"Friction bearing was developed for the belt pulley, so that when the clutch was pulled out, the pulley sleeve would not have a tendency to freeze on the crankshaft, as was the case with the earlier designs.

"The rear axle and the wheel hub were changed from taper single key construction to 6-B spline.

"The earlier prototypes had a Madison-Kipp fresh oiling system. This came about from the fact that Hart-Parr was shouting to the housetops about the fresh oiling system on their tractor engine. By 1922, however, the tractor had a pressure lubricating system which forced oil under pressure to the main and connecting rod bearings through a drilled crank.

"The shaft to drive the cooling fan was of two-piece design with an electric starter Bendix spring to pick up the shock from acceleration of the engine. This did not work out very well, however, and a friction-type clutch was developed and mounted on the front end of the driveshaft.

"The first drive disk had to be made out of brass because the original cast iron ones would stick and tear up the drive.

One of the second series of six prototypes built during 1921. *Deere Archives*

"Upon investigation, it was found that Ford had no objections to the mounting of the front axle underneath the cylinder block by attaching the axle carrying and pivoting castings to the block. This provided simple means of supporting the front end of the tractor and also of providing radius rods. It took a lot of the cost out of the job.

"Because of the fact that the front axle could be mounted to the cylinder block, it also provided low cost means of supporting the radiator in that the radiator was supported with the two water outlet elbows, one on each side of the cylinder head. This returned the cooling water to the hottest part of the hottest part of the engine.

"In connection with developing the anti-friction bearings in the pulley sleeve, the direction that the hand clutch lever as operated from changed from pulling it to engage to pushing it to engage—so that, when you disengaged it you pulled back on it.

"The Thermo-Syphon cooling system was also developed during the period of 1919 to 1922." (Brown wrote that it was introduced following the summer of 1921.)

The experimental work carried out in 1921 and 1922 occurred during one of the worst economic downturns of this century. Brown wrote that in March 1921 the tractor plant was shut down, the Marseilles plant nearly so, and the plow works operated only three days a week. On August 30 he wrote, "The business depression is awful. The shops have all been shut down for a long time and there is not prospect of any immediate improvement."

Despite economic trials, work on the Model D prototype continued, but not without some compromises. The choice of engine for the new tractor was, perhaps, the most significant. As I wrote in *John Deere General-Purpose Tractors*

Above and right: These photos of 1922 reveal a prototype closely resembling the first production machines. This tractor was most likely one of eleven or twelve built in the third series of 1921. *Deere Archives*

1928-1953, "Deere's choice of the two-cylinder engine was largely dictated by cost. In 1922, Leon Clausen, who preceded Wiman as Director of Manufacturing, sought board approval to develop a four-cylinder engine for use in what became the Model D. However, because sales of the Waterloo Boy tractor were so poor in 1921 (seventy-nine units sold) and 1922 (with losses exceeding $400,000), the company held a large inventory of two-cylinder engines. The board rejected Clausen's request for a four-cylinder engine on the basis that the company was not in a position to spend the money. Clausen had no choice but to make use of the existing inventory of engines.

"As Broehl stated in *John Deere's Company*, the board made the decision to continue with the two-cylinder engine 'not on technical grounds but on the basis of [a] short-term embarrassed financial situation.'"

The economic situation was so bad that on July 31, 1923, two months after the first production tractor was built, the very question of building the Model D came up for a vote in front of the Board of Directors. While the tractor survived, it is most interesting to note that both Theo Brown and Charles Wiman voted that the company abandon the tractor business entirely.

The directors' apprehension over the company's future as a manufacturer of tractors was soon overcome. The overall tractor market grew a healthy 15 percent in 1923. While the market was still dominated by Fordson, the Model D had been well received. The company's commitment to building tractors was secure, and by the end of 1924 the board agreed to produce 3,800 of the Model D for the following year.

One of the first 50 Model Ds, distinguished by its straight steering rod, 26in spoke flywheel, fabricated front axle, and radiator style. *Deere Archives*

Chapter 6

General Purpose Tractors

Model C

International Harvester's Farmall, the industry's first row-crop or general purpose tractor, was introduced in the spring of 1924. Deere responded to Farmall by initiating work on its own all-purpose design. Charles Deere Wiman, then Director of Manufacturing, gave Theo Brown responsibility for development of what was originally designated the "All-Crop" tractor.

Work on the All-Crop was directed along two lines: the design of the tractor and the design of an effective cultivator. Brown made his first sketches in mid-June 1925, at which point he wrote, "It seems to me that if a tractor can be made satisfactory for both plowing and cultivating that the other farm operations should be of easier solution. The idea I have is to have the tractor simply straddle one row. . ." By October, Brown, working with other Deere engineers, had decided that the All-Crop and its cultivator should be configured in a way that would permit three-row operations. The first full-scale model of the proposed All-Crop was ready by mid-October. Meanwhile, the three-row cultivating concept was tested in the field on a Fordson tractor.

Brown's drawing of the All-Crop with three-row cultivator, dated October 9, 1925. *Theo Brown's Journals, Worcester Polytechnic Institute*

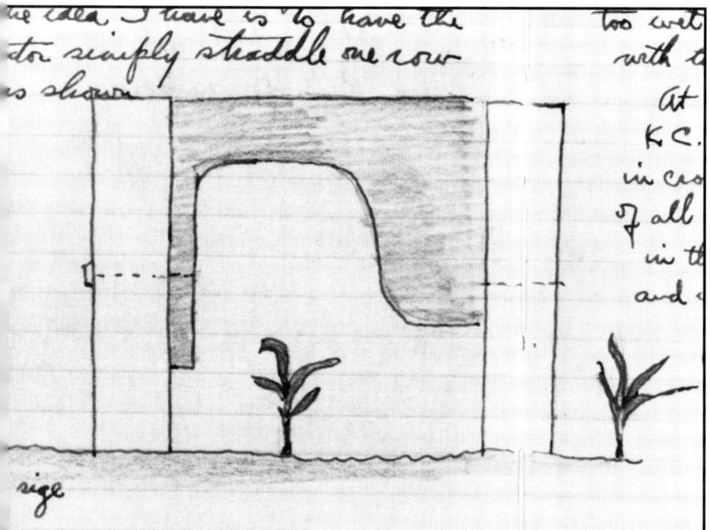

Theo Brown's first sketch related to development of a universal tractor illustrated his idea that the tractor should "simply straddle one row. . ." *Theo Brown's Journals, Worcester Polytechnic Institute*

Although there was some resistance to the three-row concept within the Deere organization, the field tests were considered a success. While Brown and his group set to work building the cultivator, H.E. McCray of the Waterloo tractor works built a second full-scale model of wood. The model was completed in December. Brown wrote, "The outfit looks good and the center of gravity not so high as we feared. The top of the radiator stands three inches higher than the Model D. . . Waterloo will try to lower the height of the tractor but it may prove impractical."

Above and below: One of the first All-Crop prototypes of 1926. In comparison to later models, its wheelbase was shorter, and its radiator was both shorter and narrower. Note the left-hand steering. *Deere Archives*

Brown recommended that clearance under the frame be reduced to 24 inches, thereby reducing the overall height of the tractor by four inches. By mid-January 1926, McCray provided new plans in which the tractor had been lowered six inches from that of the second model. On January 19, 1926, the Board of Directors authorized $25,000 for the construction of the first prototype.

The first All-Crop prototype was in the field on May 19, 1926, and was operated with both cultivator and a No. 40 plow. The prototype incorporated Brown's design for a mechanical power lift, driven off the tractor's power take-off. It and the tractor performed well. However, the cultivator was deemed too heavy. Modified by early July, the outfit was back in the field working against a Farmall purchased by Deere for comparative purposes.

One of the first three prototypes, as rebuilt for 1927. Note the narrower, taller radiator, and its experimental five-row cultivator. *Deere Archives*

By July, three prototype tractors had been built. In August, a conference of the engineers confirmed changes that included extending the wheelbase, moving the radiator forward, and reducing the steering gear ratio. After this meeting, Brown concluded that the design was "about right." The three prototypes were rebuilt by December, and in January 1927 the decision was made to build twenty-five tractors for sale. In April, the Board of Directors appropriated $875,000 to put the tractor, now the Model C, into production.

Further testing and modification of the Model C was carried out through the course of 1927. Concern over the tractor's instability on hillsides led to further modifications to the radiator and a redesign of the front axle, lowering the pivot point and adding weight.

For a while, the entire project seemed in question. The Model C was noticeably short of horsepower, a fact that created anxiety among Deere management. The fear was that the Model C would not meet expectations as a plowing tractor. On October 14, Brown wrote, "Wiman seems worried that the Model C will not do everything that the D will do. I hope we move cautiously in suggesting any change."

Four days later he wrote, "We are all much concerned over the Model C, that is to make it right for plowing. I certainly hope it can be done and save the all-purpose feature."

Support for a general purpose tractor was not unanimous, even at this late stage of its development. Brown wrote that the Minneapolis branch favored a two-plow tractor, a smaller version of the Model D. On October 19, Charles Wiman telephoned Brown asking him to design a two-bottom tractor "simply as a plow." By the end of the month, however, Brown wrote that Wiman had relaxed his concerns enough to affirm that work should continue on the Model C "unhampered by the thought of a new small 'D'."

Work did continue: the tractor's power take-off clutch was redesigned; and most notably, a new cylinder head was developed, which, Brown wrote in January 1928, was expected to boost the Model C's performance to "3/4 of the new Model D capacity."

That month, the decision was made to adopt the Model C tractor and to rebuild all units manufactured prior to January 24, 1928. Production of the Model C ended in April 1928. By June, the tractor was redesignated the Model GP and full-scale production followed beginning in August.

GP-Tricycle

There are indications that a few units of the Model C were built with a tricycle front-end, however, the original version of the Model GP was available with a standard front axle only.

In mid-October 1927, despite the eleventh hour

jitters that called into question the need for a general purpose tractor, Theo Brown sketched an idea for a tricycle front two-row cultivating tractor. By November 21, Brown had a prototype in the field, and on December 31 it was loaded onto a rail car and shipped to California for winter trials. A second prototype would follow in mid-January 1928.

A four-row cultivator was built and tested with the tricycle prototype. Brown wrote that it performed well, and the decision was taken in June to build 12 tricycle tractors by September for shipment to Corpus Christi, Texas. In November, production of another fifty was allocated.

Of the approximately seventy GP-Tricycles built, four to six units were built with 68in rear tread width, for use in potato rows. Three of these units were shipped to Maine. The remaining units were essentially the Standard GP, but with a tricycle front. Most of these tractors were shipped to Texas, presumably for sale to cotton growers.

Production of the GP-Tricycle was suspended, following introduction of the General Purpose Wide-Tread. Prized by collectors, only one example of the GP-Tricycle is known to exist.

General Purpose Wide Tread

The Standard GP and its three-row design were better accepted in the North than in the South, where cotton growers preferred the tricycle-front, wide-tread rear axle, and the two- or four-row configuration of the Farmall and its implements. Farmers complained that the GP's broad hood and radiator and its standard front axle hampered forward visibility when cultivating. (Poor visibility alone may have made the GP's three-row cultivator impractical.)

Whatever the reasons for its poor sales, Deere realized that the GP would have to be redesigned to better compete with the Farmall. In July 1929, Deere began manufacture of the "General Purpose Wide-Tread." With its two-wheel tricycle front, wide-tread rear axle, and two- and four-row cultivators, the Wide-Tread more closely resembled the Farmall.

Rear view of the 1927 prototype, briefly designated the Powerfarmer. Note the steering position is on-center. *Deere Archives*

Above and next page: Front and rear views of the 1927 Powerfarmer, this example with left-hand steering. *Deere Archives*

General Purpose Orchard Tractor

In 1929, Lindeman Power and Equipment, a Deere dealer in Yakima, Washington, modified a Standard Model GP in a manner that lowered the unit to a height suited for clearance in orchards.

Lindeman fitted a new front axle and modified quill castings, which lowered the tractor's overall height by seven inches. Once completed, the tractor was demonstrated for officials of Deere and Company. Impressed by Lindeman's concept, Deere agreed to build its own prototypes incorporating features of the Lindeman design.

Deere began with six General Purpose Wide-Tread units from stock (serial numbers 402115, 402128, 402138, 402147, 402245, and 402250) and modified them to suit. The units featured the 6inx6in engine with crossover manifold.

To reduce its height, Deere's engineers employed a lower front axle and reversed the rear axle housing, which thereby lowered the tractor's main frame. This ingenious modification permit-

ted the operator platform to be dropped behind the housing and positioned the operator between the rear wheels and fenders.

Further, the tractor's intake stack was eliminated and the exhaust stack and muffler were positioned horizontally and parallel to the tractor frame. Measured at the radiator cap, the overall height of the Orchard version was reduced by 8 1/2in from that of the Wide-Tread, and by seven inches from that of the Standard model.

Between July and September 1930, the six prototypes were shipped to Washington and California for evaluation. In February 1931, the company issued the decision to proceed with manufacture of the General Purpose Orchard Tractor, and production began on April 2, 1931.

Further Improvements to the GPWT

While the new GPWT offered an improved view for cultivating, it was still not ideal. Further, its steering had also come under criticism, as the front wheels tended to "whip" due to the tractor's somewhat indirect steering system. Offset to the right of the tractor, it employed a shaft and worm that ran from the steering wheel to the worm gear housing, mounted at the top and right-hand side of the transmission case.

A link dropped from the worm gear, and was connected by a ball-joint to the drag link that ran at the right-hand (pulley) side of the tractor, level with its frame. The drag link was connected to a steering arm, which in turn steered the yoke and spindle.

A 1927 Model C, serial no. 101. The steering wheel appears to be positioned higher than that of the Powerfarmer. *Deere Archives*

By May 1931, a new prototype GPWT had been built incorporating an overhead steering system similar to that of the Farmall, but employing a rising worm and worm gear instead of beveled gears. The overhead steering system, which was set in the center of the tractor, employed a shaft and worm that ran from the steering wheel, through a support mount, and over the hood of the tractor to a bolster or pedestal mounted in front of the radiator. The worm gear housing was mounted to the top of the bolster, with its gearing connected directly to the yoke and spindle.

Additional changes were made to further improve the operator's line of vision. The fuel tank was tapered, the seat raised and brought forward to help see over the pulley, and the air cleaner and muffler were moved under the hood. The prototype was tested through the fall of 1931, before the changes were adopted for production in February 1932.

GP Argentine Special

The General Purpose Wide-Tread offered fixed 76in rear tread width, suitable for working in rows up to 38in wide. In early May 1930, a Deere executive returned from a visit to Argentina, where he concluded that there was a substantial market for a general purpose tractor suited to 28in rows. By June 20, Deere had modified at least one GPWT for export to Argentina. This "Argentine Special" featured 56in rear tread width and four-row cultivator suited for narrow corn rows.

Model GP Series "P": Potato Growers' Special

Of the first twenty-three General Purpose Tricycles built, at least four—and perhaps six—featured 68in rear tread width, suited for use in potato rows. In November 1929, as a result of the experience gained with these units, Deere elected to build a special "potato" tractor, designated the General Purpose Wide-Tread Series "P."

As production of the GP-Tricycle had ended, the new tractor was based on the General Purpose Wide-Tread. To permit a shortened 68 inch wheel tread, the GPWT's right and left quills were shortened and the differential shaft altered.

Prior to production of the Series P, the factory estimated annual demand at 150 units. In fact, a total of 203 units were built in two production runs, before production was terminated. The first run commenced January 30, 1930, with serial no. P5000 and ended February 27, with serial no. P5149. By the end of the first production run, Deere had introduced the 6inx6in engine. The company was also prepared to introduce an offset rear wheel for the Wide-Tread, that effectively narrowed its rear tread width to 68in. This new wheel made the Series P redundant.

According to *Two-Cylinder*, "to keep all the 'P' tractors alike, and to satisfy existing orders prior to

Rear view of the Model C, serial no. 101. *Deere Archives*

A Model C disassembled for inspection. March 1928.
Theo Brown's Journals, Worcester Polytechnic Institute

offering [the] special potato wheel for the regular Wide-Tread," Deere converted an existing inventory of 73 5flin bore Standard tractors to Series P specifications.

This final run of Series P tractors commenced on July 25, 1930, with serial no. P5150 and was ended August 19 with serial no. P5202.

Model A Prototypes: Modifications to the Model GPWT, and the Models FX, GX, AA-1, and AA-3

The General Purpose Model A was introduced in 1934 and was built through 1952. With more than 300,000 units sold, the Model A was one of Deere's best selling series of tractors, second only to the Model B in popularity. In this section, we explore the development of the early or "unstyled" Model A.

The Model GP was not a particular success for Deere. Its sales lagged far behind those of Farmall, despite the improvements incorporated in the GPWT. By December 1930, discussions of a new general purpose design were underway. At this early stage, three significant features of the new model had already been defined: hydraulic power lift; front axle adjustment; and variable rear tread width. While the first Model A prototypes did not appear before November 1931, work on the hydraulic lift and adjustable rear tread width features was begun much earlier.

Hydraulic Lift

Work on an hydraulic implement lift began in October 1929, and was intended for introduction on the Model GPWT. The first working prototype entered field trials in January 1930. Fitted with a 7in cylinder, it lifted 8,000lbs at 200lbs pressure. Considered too slow, the unit was redesigned with a smaller cylinder and tested extensively throughout 1930. By March 1931, fifteen hydraulic lift units were built for more extensive trials. By this date, however, planning for a new general purpose tractor, the future Model A, had begun. As the hydraulic lift system was not yet perfected, it was decided to continue its development but postpone its introduction until the proposed new tractor was built.

Adjustable Rear Tread

The earliest work on adjustable or variable rear tread was initiated in December 1930. The first prototype was built in March 1931 by modifying a Model GPWT. Designed to work with the hy-

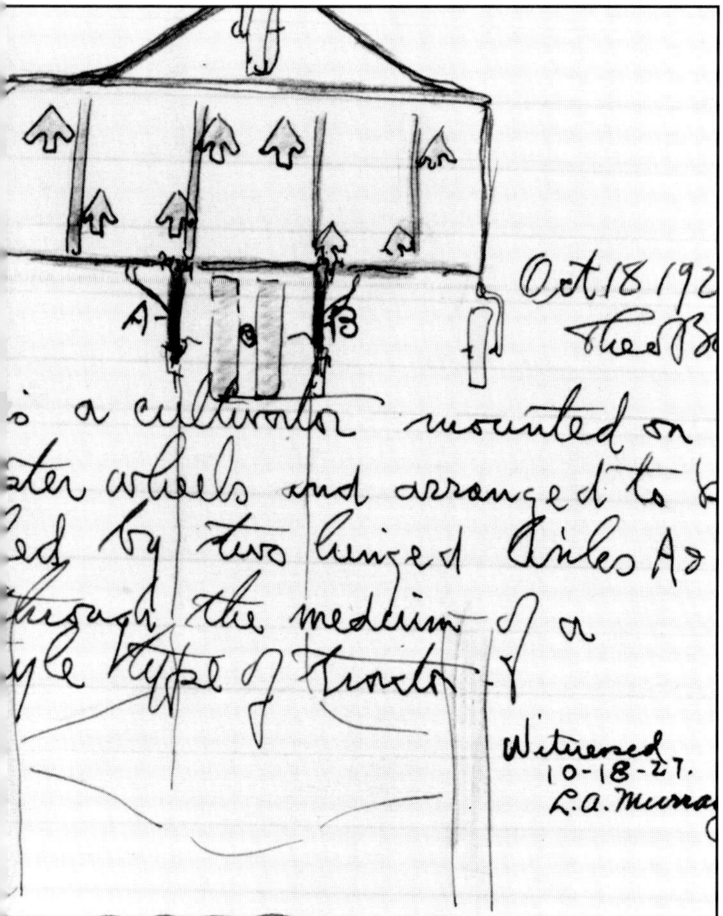

Theo Brown's first sketch of a two-row, tricycle-type GP tractor. *Theo Brown's Journals, Worcester Polytechnic Institute*

draulic power lift, tread width was changed by running the tractor along the ground and using the power from the power lift to rock a rockshaft and ratchet the gears to work the tread in or out. Holding the power lift in continuously widened or narrowed the tread. This early system, as designed by Theo Brown, was not favored by engineers in Waterloo.

Models FX and GX

In January 1931, Charles Wiman appointed Theo Brown, along with three other men to the "Power Farming Committee," to define the new general purpose tractor. The group met many times over the next few years, as the Model A evolved.

At their first meeting on February 3, 1931, the committee discussed the need for a tractor offering high clearance and the ability to handle crops in 20 to 42in rows. There was a consensus that an engine that developed 30 belt horsepower was required, and that both two- and four-cylinder motors should be considered.

By September 1931, the new tractor was designated as the FX. The first full-scale model was begun on September 14 and completed by end of that month. The first prototype was built by the end of November. It weighed 4,600lbs, and its two-cylinder engine generated 29 belt horsepower.

The horsepower requirement for the FX was a major consideration. In December 1931, the committee recommended building two new general purpose tractors; one, a 36hp model; the other, a 24hp tractor built along the same lines as the larger machine.

When considering both 24 and 36hp tractors, the committee was no doubt influenced by the competition Deere faced from Farmall. Over 100,000 Farmalls had been sold by mid-1930. The new Farmall F-30, a three-plow, 30hp companion to the original Farmall was also in production. New two- and three-plow general purpose tractors from Deere would certainly be a match for their foremost competitor. However, by January 1932, Deere dropped the notion of building a 36hp tractor. The plan to focus on just one new tractor was very likely a function of the poor state of sales and profits at Deere during the Depression. Sales in 1931 were off by more than 50 percent from 1930 and by nearly 65 percent from 1929. The company had lost $1 million dollars in 1931. In February 1932, Theo Brown wrote, "Business is depressing. There literally is none . . . and at the shop we are trimming our expenses to the very minimum. It is hard."

Despite cutbacks, work continued on development of the FX. Attention was focused on cutting the weight of the tractor, perfecting the hydraulic lift, and increasing front axle clearance. In March 1932, an FX prototype was sent to the Kansas City branch for field testing. Another FX tractor was tested extensively at Deere's experimental farm near Moline.

By May 1932, engineers at Waterloo had revised the FX to the point that it was given a new designation, GX. Theo Brown wrote that the GX featured 27in clearance under the front axle. The tractor's weight had been trimmed to 3,700lbs. By early June, approval was granted to build two prototypes for testing.

The first GX prototypes were built and running by mid-September. The new model weighed 3,350lbs. Its engine generated 24.75 belt horsepower. Although no record remains of the number of GX tractors built, seven units were shipped to Phoenix

GP Tricycle, photographed in September 1928. *Deere Archives*

and Tucson in December 1932 for field tests.

By February 1933, the decision was made to build ten tractors for sale, and these were given the designation "AA." A provision was made for two versions, a four-speed model and a three-speed model. Production records indicate that only eight were built: six of the four-speed Model AA-1, and two of the three-speed Model AA-3.

By June 1933, the choice of transmission was the only decision pending before production could begin. On June 14, both versions of the AA were demonstrated for a group of company officials that included Charles Wiman, Frank Silloway, and Theo Brown. That same day, the decision was made to build the four-speed version only. On July 25, 1933, the Board of Directors approved $463,756 to begin manufacture of fifty units a day as of December 1.

Production of the Model A was delayed, however. In mid-January 1934, Theo Brown wrote that there were problems with engine cooling and oil consumption. The problems were overcome and on March 19, 1934, the first Model A came off the line.

Model B Prototype: Model HX

Although the Power Farming Committee had rejected the idea of building a three-plow tractor in early 1932, by July 1932 they were meeting to formulate requirements for a smaller companion to the Model GX.

Above and below: Two GP Tricycles. Tractor with lower steering wheel position from June 1928, other unit photographed in August. *Deere Archives*

Above and right: GPWT modified with overhead steering. *Deere Archives*

The committee established the following requirements for the new tractor: the ability to handle all row widths from 20 to 42in, with the exception of 22in rows, by 2in increments; wheelbase not less than the maximum tread, a minimum of 80in; stability equal to that of the 1932 GP Wide Tread; vertical clearance of not less than 22in; power sufficient to handle one 16in plow bottom, or 15hp; three-speed transmission; belt pulley; and power take-off.

By mid-January 1933, the first drawings and models of the proposed "HX" tractor had been completed. The tractor was designed to be as near 65 percent of the cost and weight of the GX as possible. Both three- and four-speed transmissions were considered.

Theo Brown wrote that it was hoped that three or four HX prototypes would be built by mid-April. In fact, seven HX tractors were built in July and August, with the first one in the field on August 29. In January 1934, it was decided to build another seven HX tractors for continued field trials. Of the fourteen tractors, two were to be sold by the Moline branch and one, equipped with power lift, was shipped to Dallas for sale. Eight tractors were held for testing at the Tractor Works, and the remaining three were shipped to Moline, St. Louis, and Dallas branches for testing.

On January 30, 1934, the Board of Directors approved production of the HX and appropriated $250,000 for start up. Further testing of the HX continued through the spring and summer. Production was scheduled to begin in September, and the first Model B came off the line on September 28.

Max Sklovsky's One-Plow Tractor: A Proposed Alternative to the Model B

In late 1932, as the first layouts of the HX tractor were being prepared, Max Sklovsky proposed an alternative design for the Power Farming Committee's consideration. The tractor featured three-wheel construction, with power delivered to the single rear drive wheel. The front wheel diameter measured 20in. The rear wheel measured 24in.

In March 1933, the committee carefully evaluated the features of the Sklovsky design versus those of the proposed HX tractor, and conducted three days of field trials. In the end, Sklovsky's tractor was rejected for a number of reasons. Chief among them were the instability of a three-wheel tractor; the fact the tractor's rear wheel was too

A limited production GP potato tractor, the Model P. *Deere Archives*

wide for work in potatoes and could not work in listed rows; the tractor could only handle a single 14in plow bottom—whereas the committee had already established that a new one-plow tractor must handle a 16in bottom; and its maximum 600lbs drawbar pull was considered inadequate.

Longer Wheelbase for the Model B

In October 1936, the decision was made to lengthen the wheelbase of the Model B by five inches. This permitted the interchange of mounted implements with the larger Model A.

The first prototype incorporating this change was built in May 1936. At least one other similar prototype was built and tested before the decision was made to make the change. The first production Model B with lengthened frame was built in June 1937.

Model G Prototype: Model KX

Once the Model B was clearly on its way, focus again shifted to building a three-plow general purpose tractor. The Model KX was developed by Deere as a competitor for the Farmall F-30, which was introduced in 1931. The KX would be introduced in May 1937 as the Model

G (it was at one point designated as the Model F) and work on its design began in late 1934. As was the case with the Model B prototype, little original work was required, as the tractor's basic features and configuration closely followed those of the Model A—it was simply designed on a larger scale.

In March 1935, Theo Brown became Chief of Implement Research. At that time, he wrote of the need to changeover the front end of the KX to provide the same horizontal and vertical implement attaching points as those of the Model A. These changes permitted the use of the same cultivator on both tractors. The final design also incorporated the same attaching points on the rear axle housing as the Model A, which permitted all mounted implements to be interchanged.

The first two KX prototypes were in the field by late August 1935; one cultivating, the other plowing. The tractor weighed 5,100lbs, but prior to production the weight was reduced by more than 500lbs. In total, eight to ten experimental tractors were built.

Perhaps the most significant design decision affecting the Model G was made following introduction of the tractor, when problems of overheating were reported during the summer of 1937. Work at improving the cooling system was executed by the Engineering Department and a larger radiator was introduced in January 1939. The problem was addressed in the field through an update program. Eventually, the tractor's block and cylinder head were redesigned to further improve cooling.

Model L Prototype: Model Y
Model 62 and Model LW

About the same time that the Model KX—the larger version of the Models A and B—was under development in Waterloo, work on a smaller companion was carried out at the Moline Wagon Works. The Model Y was designed for the small farm and truck garden.

Approximately twenty-five Model Y tractors were built in 1936. The Y featured a welded frame, Novo Model C-66 or Hercules NXA 8hp, two-cylinder vertical engine, foot-operated clutch, and a Ford Model A three-speed transmission and steering gear. The steering wheel was offset to the left side, which provided an unobstructed view to the right when cultivating.

The Model Y was offered for sale in 1936, as were a specially suited planter and cultivator. Of the units built, an unknown number were sold, all were recalled, and none have survived. Their final disposition is unknown, although the chassis of one unit served as the basis for Theo Brown's "One-Row Cultivating Tractor," the earliest example of his Model 101.

The Model Y was succeeded in 1937 by the Model 62. Fitted with the Hercules NXA 8hp, two-cylinder vertical engine and a three-speed transmission, its basic configuration closely followed that of the Model Y. Unlike the Model Y, however, which was considered an experimental tractor, the Model 62 was a limited production tractor. Approximately seventy-eight units were manufactured before full-scale production as the Model L was begun in August 1937.

The Model L was fitted with a standard 40in tread width front axle. While it proved popular, there was a demand for a small tricycle-type tractor with widespread rear wheels. In 1938, one example of such a tractor, the Model LW, was built by modifying a Model L. An engineer involved in the project cited the disruptions of World War II as the reason the tractor was never developed to the point of production.

A Model GPWT modified with 56in rear tread width and fitted with a four-row cultivator, as exported to Argentina in 1930. *Theo Brown's Journals, Worcester Polytechnic Institute*

Model H Prototype: Model OX

By the time production of the Model L began, Waterloo engineers had designed their own version of a one-plow, two-row general purpose tractor with features that more closely paralleled those of the Models A, B, and G. Designated the Model OX, the tricycle-type GP tractor was presented to the board in November 1937. The first prototype was in the field by February 1938, and by mid-year six additional prototypes were built.

The first series of six prototypes were soon followed by a second series of six. While there may well have been differences in their mechanics, the two versions of the OX prototype are best distin-

The first prototype hydraulic power lift, a feature that was introduced on the Model A, as fitted to a Model GP in January 1930. *Theo Brown's Journals, Worcester Polytechnic Institute*

A Model GPWT fitted with hydraulic power lift. October 1930. *Theo Brown's Journals, Worcester Polytechnic Institute*

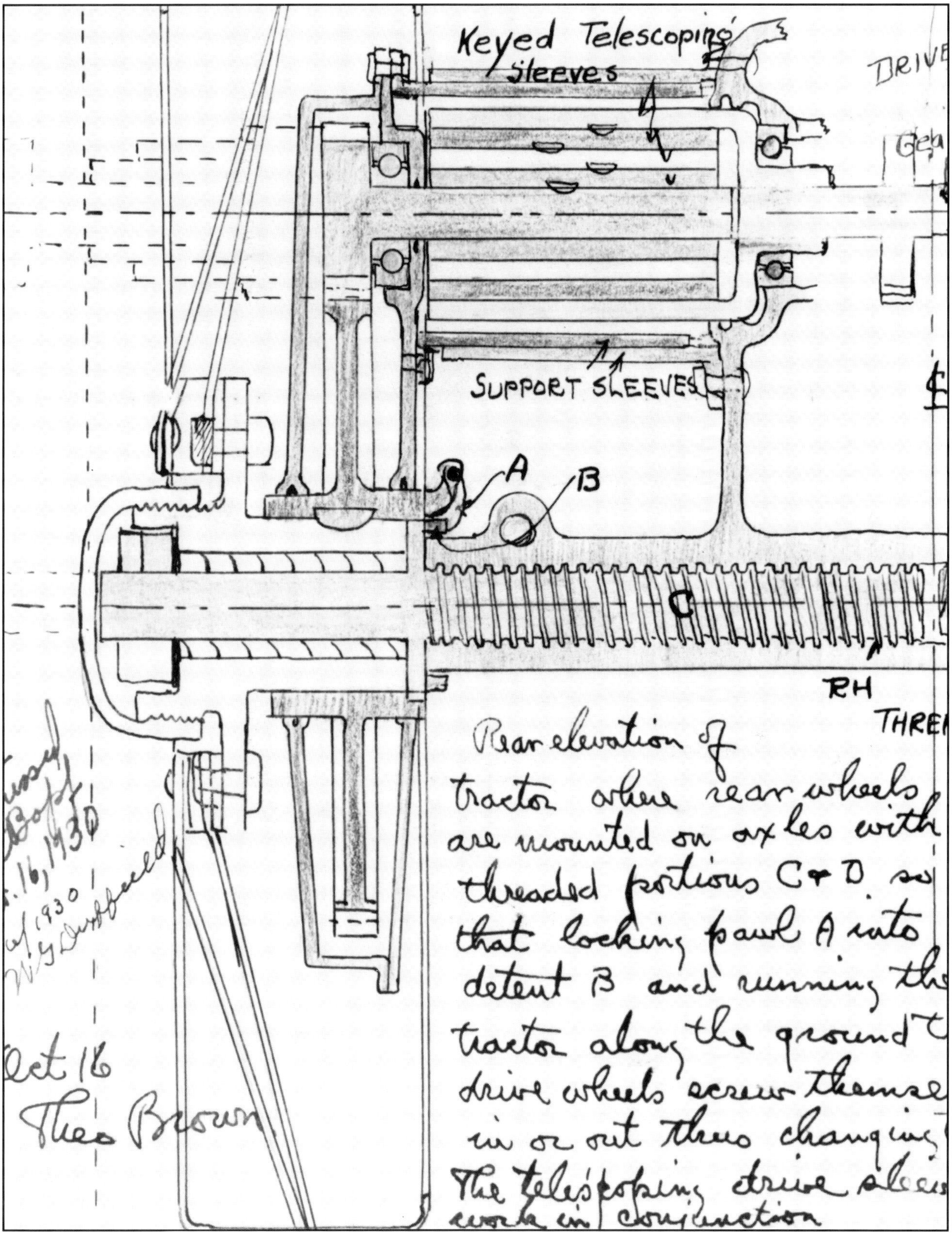

Theo Brown's October 1930 drawing of a telescoping rear axle, his first effort at developing an adjustable rear tread. While not adopted in this form, the Model A became Deere's first tractor to offer adjustable rear tread. *Theo Brown's Journals, Worcester Polytechnic Institute*

Above and right: Model GP modified with telescoping driveshafts and notched support shafts ratcheted in or out by power supplied by the tractor's power lift. *Deere Archives*

guished by their individual styling. The early OX featured an exposed radiator and backswept steering pedestal, as well as upright intake and exhaust pipes. Its seat was bolted to a forged leaf spring mounted via a bracket bolted to the rear axle housing cover and a bolt through the top of the transmission housing.

The later OX assumed the Henry Dreyfuss "styled" sheet metal adopted by the Models A and B. The upright intake pipe was eliminated, replaced by a screened vent at the left side of the hood. A permanent platform and raised steering and seat position were also adopted. The seat itself featured coil spring suspension.

The decision to build the Model H was issued in September 1938, with manufacture of the first production tractor following at the end of October. The Model H was brought from prototype to production in less than twelve months, a decision very likely precipitated by competitive pressures. In December 1937, Allis Chalmers introduced the Model B, the first true row-crop tractor to sell for less than $500. Praised by *Farm Implement News* as a tractor that would "do anything on a small farm and cost not much more than a team and far less to operate," the Allis Chalmers B sold at an impressive rate. It clearly demonstrated that the demand for small general purpose tractors was as great as that for larger general purpose tractors. While the Deere Model H was priced higher than the Allis Chalmers B, it too advanced the industry's goal of completely replacing animal power.

The rapid development of the OX may well have resulted in the failure to discover a weakness in the engine's cast alloy crankshaft. Although no recorded evidence of crankshaft failure remains,

almost all of the first 104 production tractors were reportedly either scrapped or fitted with new forged steel crankshafts and assigned new serial numbers (the new crankshaft going into production tractors beginning in January 1939).

Of the twelve or so OX prototypes built in 1938, two were reportedly maintained at the factory until the late 1940s, but all are now presumed to have been scrapped.

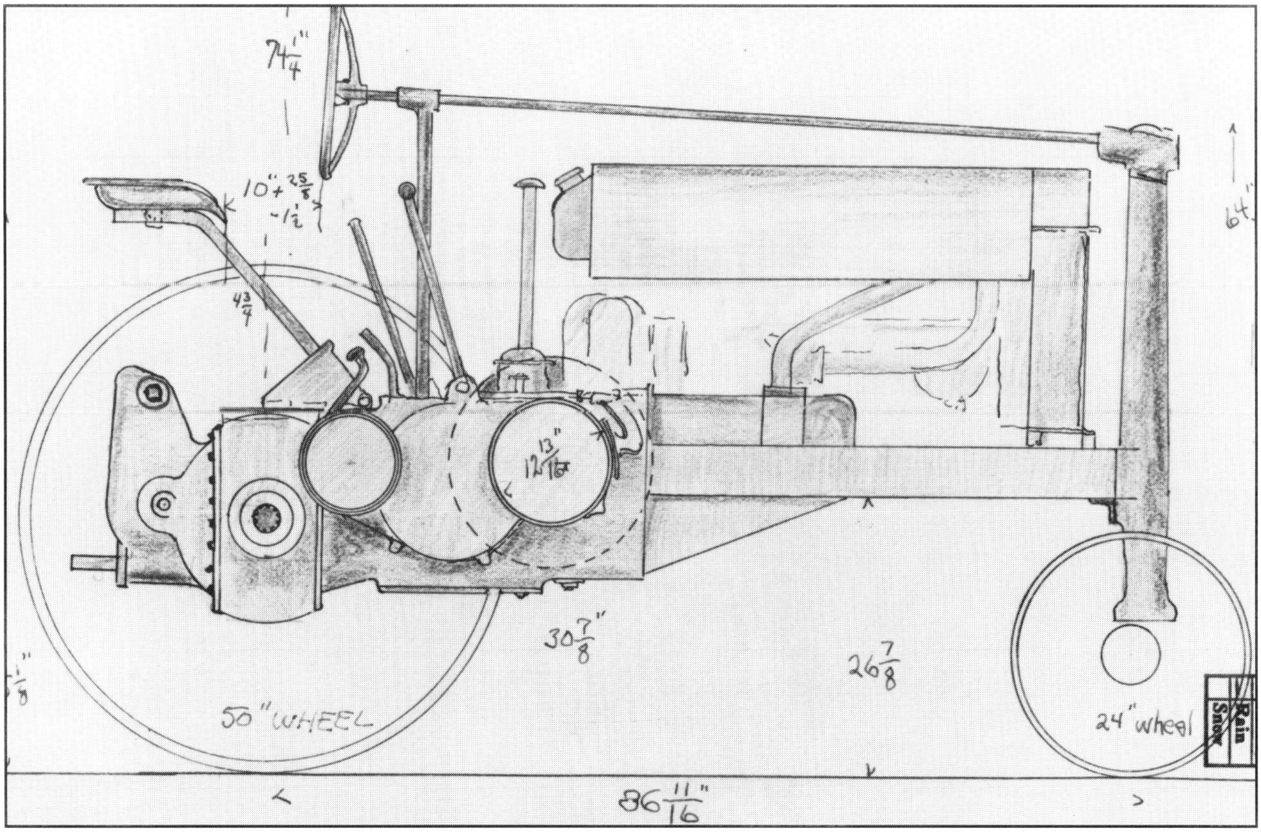

A drawing of the GX prototype executed by Theo Brown on June 25, 1932. *Theo Brown's Journals, Worcester Polytechnic Institute*

Above and left bottom: Two photos taken in late October 1932 of GX prototypes. Note the differences: one features clamps on the rear wheel hubs; the same tractor has a universal joint in the steering shaft. Also note the water pipe running to the engine head, rather than to the block as with the production engine. *Theo Brown's Journals, Worcester Polytechnic Institute*

A Model AA, most likely an AA-1, the four-speed prototype of the Model A, in July 1933. *Deere Archives*

Above and left bottom: Two views of a scale model of the AA photographed in October 1933. *Deere Archives*

Below and right: Front and rear views of an AA prototype. Note absence of the John Deere name on the casting above the radiator and lack of decals. *Deere Archives*

Model HX, prototype of the Model B, in a face off with a Farmall F-12, during a September 1933 field trial. *Theo Brown's Journals, Worcester Polytechnic Institute*

Rear views of the Model HX and Farmall F-12. With its three-speed transmission and four-cylinder engine, the F-12 barely out-pulled the HX with its two cylinders and four speeds. *Theo Brown's Journals, Worcester Polytechnic Institute*

Flywheel side view of the HX photographed near the factory on September 21, 1933. *Theo Brown's Journals, Worcester Polytechnic Institute*

A later version of the HX, with integral 16in plow, under test on April 20, 1934, near Moline. *Theo Brown's Journals, Worcester Polytechnic Institute*

Max Sklovsky at the wheel of his One-Plow tractor. March 1933. *Deere Archives*

Above and right: The Sklovsky One-Plow was proposed by Max Sklovsky as an alternative to the Model HK prototype (Model B). It did not fare well in a comparison of features. *Deere Archives*

The Model B underwent a number of changes through the course of production. Here is a May 1946 prototype of what became the "late-styled" Model B. Distinguished by its pressed steel frame and slightly larger engine, production began in February 1947. *Deere Archives*

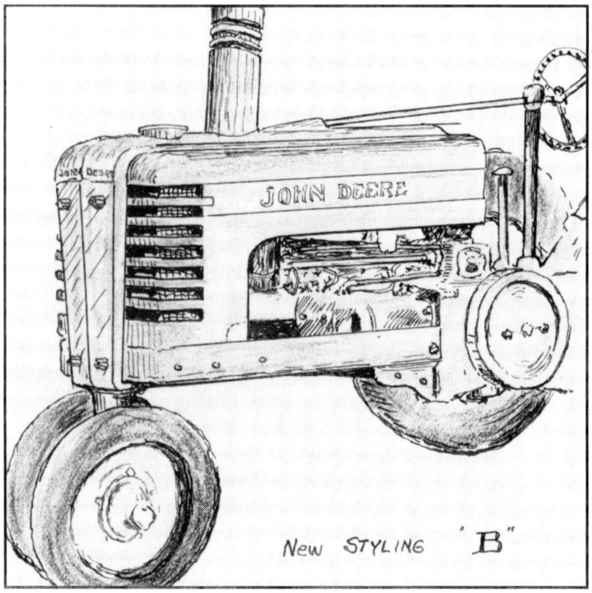

Theo Brown captured his impression of the Henry Dreyfuss styling for the Model B from blueprints he received on March 4, 1938. *Theo Brown's Journals, Worcester Polytechnic Institute*

In May 1936, this Model B was constructed with its frame lengthened by 5in. According to Theo Brown, only one such tractor was built for testing. The prototype is shown with a Model 290 cultivator, designed for the larger Model A. The longer frame was adopted, but not before June 1937. *Theo Brown's Journals, Worcester Polytechnic Institute*

JOHN DEERE
JOHN DEERE

Studio photographs of the Model Y, one of approximately twenty-five tractors built in 1936. *Deere Archives*

A Model Y in the field. November 1936. *Deere Archives*

The Model 62, a limited production tractor, followed the Model Y. The John Deere "JD" logo cast below the radiator distinguished the Model 62 from its successor, the Model L. *Deere Archives*

Left and below: Overhead and rear views of a prototype Model 62. *Deere Archives*

A Model 62 with ground-driven mower. June 1937.
Deere Archives

Above and right: Opposite side views of the unstyled, or first series, of Model OX prototype photographed in May 1938. *Deere Archives*

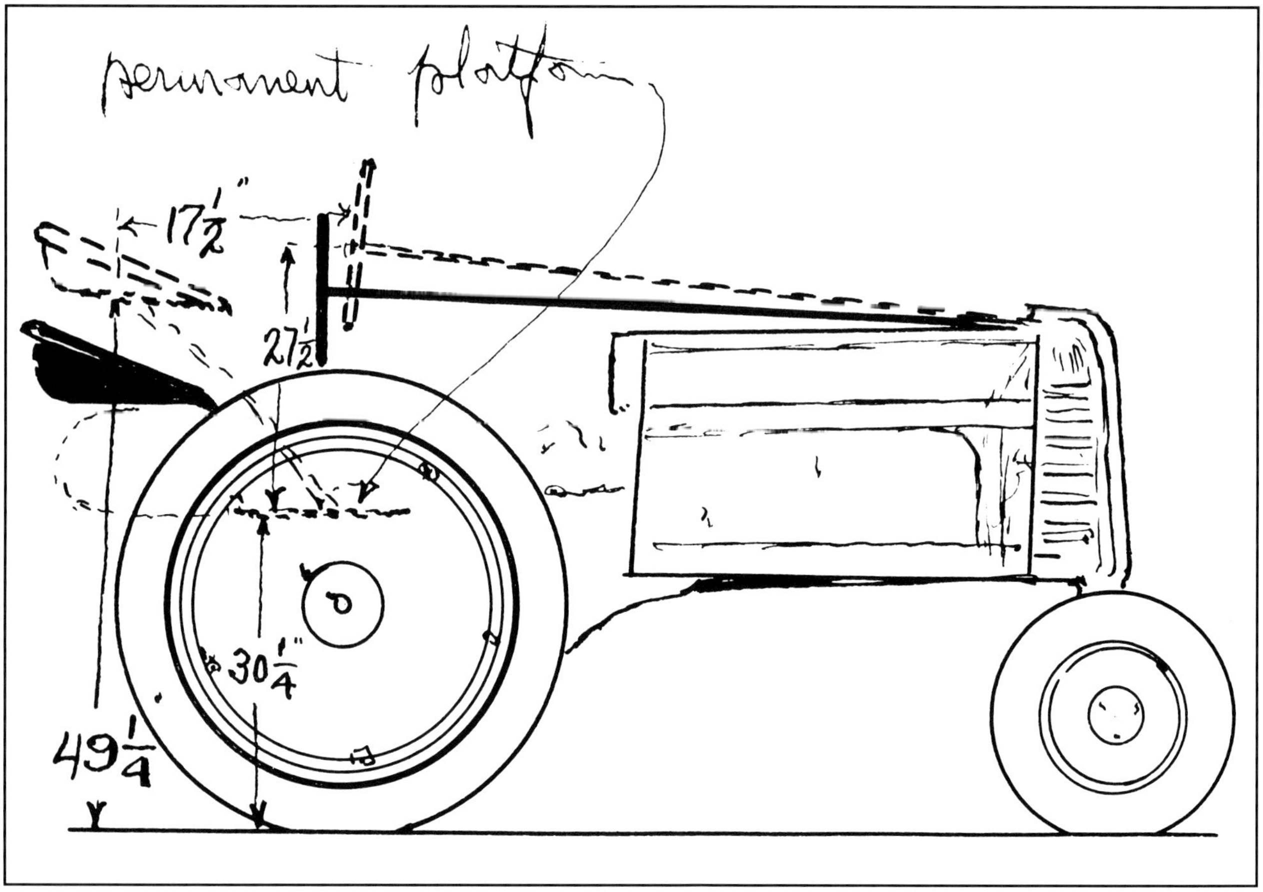

Theo Brown recommended the addition of a permanent platform to the OX, and sketched suggested subsequent changes to steering and seating positions on March 24, 1938. Note that he incorporated the Dreyfuss styling changes. *Theo Brown's Journals, Worcester Polytechnic Institute*

An example of the second series of OX prototypes. Note the screened air intake vent on the hood to the left of the John Deere decal. *Deere Archives*

Above and left: Rear views of the styled and unstyled versions of the OX. The addition of a platform and changes to the steering and seat suspension denote the styled tractor. *Deere Archives*

Chapter 7

Post-World War II Tractors

Model M Prototype: Model 69

The Model L, its heftier sibling the LA, and the Model H were all rated as one-plow tractors and were marketed as primary and utility tractors. The potential for small tractors was quite large in the late 1930s and 40s, but the marketplace was also crowded by some very popular competitors: Allis-Chalmers Models B and C, Case V and VA Series, and Farmall Models A and B.

Although Deere was occupied with war work and faced manpower shortages beginning in 1942, the experimental departments continued their work. While I could not determine exactly when the Model 69 was conceived, it was certainly prior to 1944. By January 1945, the decision to put the tractor into production had already been made, but awaited construction of the new Dubuque plant.

The most distinctive feature of the Model 69 was its upright, two-cylinder motor. While no record of the first prototype Model 69 remains, prototype No. 2 was tested exhaustively in 1944, operating more than 2,400 hours in the field. Plowing, disking, loading hay, pulling an eight-foot straight drag, and driving a 10A combine, No. 2 consumed 3,576 gallons of fuel and seven quarts of motor oil. The following season it was joined by prototypes No. 3 and 4, and was tested another 1,500 hours before being completely disassembled and inspected.

The 69 gained the designation Model M in 1946 and twenty-three prototypes, including XM-17, the first crawler prototype, were tested in field trials throughout the U.S. The Model M was introduced in 1947 as a general purpose utility tractor supplanting both the Model LA and H. It was initially offered with a fixed or standard front axle only. However, plans were already in the works for tricycle and adjustable front ends, as well as an industrial version.

A Model 69 cultivating soybeans near Moline in July 1944. *Deere Archives*

Model 69 with a Model 101 corn picker in November 1944. Even at this early date, the tractor's design was firmly established. *Deere Archives*

The first adjustable front axle was tested in Lansing, Michigan, in July 1947, fitted to production tractor no. M10598. Adjustable twin and single front wheel axles were tested on XMT-01 and XMT-02 that same season. Testing of XMT and XMC prototypes continued through the 1948 season, before production of the MT and MC was begun in late December.

MX: Deere's First Diesel

In 1931, Caterpillar Tractor Co. became the first American manufacturer to offer a diesel tractor. Within four years, Caterpillar had built 10,000 diesel engines, and both Cletrac and International Harvester also offered diesel-powered track machines.

It was most likely that the introduction of first diesel wheel tractor, the WD-40, motivated Deere to begin experimentation of its own. Deere began its work with diesel engines in late 1935. Principal experiments focused on the design of a starting system which could overcome the inherent problems of starting a diesel engine under cold weather conditions; an optimal design for the combustion chamber; and the need for an enhanced fuel filtration system.

After trying more than one starting system, Deere settled on the use of a two-cylinder gasoline pony engine, the exhaust of which also served to warm the diesel engine. Working with its vendor, Deere developed a highly reliable fuel filtration system which overcame the potential of fouling the fuel pump.

Deere built its first diesel tractor prototype as early as 1938. It was undoubtedly an adaptation of the Model D, although no photographic evidence

Model 69 No. 2, one of four units tested in 1944-45, disassembled for inspection at the Laredo, Texas, test facility. *Deere Archives*

Model 69 No. 4 photographed in September 1945. This tractor was taken on a twenty-one state tour that year. *Deere Archives*

of such remains. By June 1939, the diesel tractor had gained the designation "Model MX," and at least one other prototype was built. A year later, an engineering order was issued that defined the specifications of what would become Deere's first diesel-powered tractor.

The MX prototype evolved through three series of designs: in 1941, eight MX prototypes were built and sent to the field for testing; followed in 1944 by five improved tractors; and in 1947 by another eight tractors. I found few details of any of the prototypes or descriptions of the field trials. An entry in Theo Brown's journal does refer to a January 1945 demonstration of the MX, as the "experimental 48hp diesel."

Rear view of Model 69 No. 3, still sporting a steel seat, as of December 1945. The Model 69 featured an adjustable steering wheel. *Deere Archives*

Presumably, the final eight prototypes closely resembled the production Model R. Their styling was executed by Henry Dreyfuss, who offered at least ten different designs for consideration, four of which were actually built before the final design was selected. The final choice was distinguished by its pleated grille made of perforated metal with small holes. Its design served the purposes of screening out all but the smallest airborne matter; its corrugated form was sufficiently strong to eliminate the need for added reinforcement, thereby increasing the open grille area and allowing a greater amount of air to pass through the radiator.

The Model R also was the first Deere tractor to offer a fully independent power take-off, testing of which was also carried out on the MX prototype. Production of the Model R was begun in February 1949.

The Numbered Series

The bold styling and mechanical characteristics of the Model R set new standards of design and performance for Deere tractors. Although the company had updated the Models A and B in 1947, boosting their horsepower, introducing pressed steel frames, and adding such goodies as remote hydraulics and standard electrics, the tractors still reflected technologies of the 1930s. While adequate, the A and B were no longer exceptional, and the freshness and strength of the Model R only served to mark them as such.

It was not only the Model R that dated the rest of the Deere line. Throughout the industry, technology advanced dramatically in the years following World War II. The three-point hitch, the torque converter transmission, power-adjust rear wheel tread, "live" power take-off and hydraulics, LP-gas options, and a revived interest in four-wheel drive highlighted the advancements witnessed between 1945 and 1950.

By 1950, plans were underway to supplant the A, B, and G with a new series of tractors devised to keep pace with the industry. The AX and BX prototypes of 1950 gave way in 1952 to the Models 60 and 50, respectively. The most characteristic features of the 60 and 50 included: a pleated radiator grille a la Model R, two-barrel carburetion, a redesigned cylinder chamber that improved combustion and boosted horsepower, 12 volt electrics, live power take-off, live hydraulics, improved steering, modified three-point hitch, a two-piece front pedestal with five interchangeable wheel assemblies, and "Quick-Change Wheel Tread." In 1953, many of the same features were incorporated into the Model 70, successor to the Model G. The 70 was Deere's first general purpose tractor to offer gasoline, all-fuel, LP-gas, and diesel options.

As Waterloo readied the 50, 60, and 70, Dubuque set to work on the Model 40, an improved version of the Model M. The first prototypes to appear were the 40XS and 40XMT, standard and narrow front tractors, respectively. Their most notable features included a true three-point hitch, live hydraulics, fluted radiator grille, and 15 percent more horsepower resulting from a boost in engine rpm.

At least one crawler prototype, XMC-12, was in the field by December 1952. Its improved traction and stability resulted from a four-track roller running gear. (The MC featured three track rollers; the production 40C would offer a choice of four or five track rollers.)

Experimental Model M No. XM-07, one of twenty-three Model M prototypes tested in 1946. *Deere Archives*

Experimental tractor XM 17, photographed near Yakima, Washington, in December 1946. The crawler version of the Model M was developed by Lindeman Power and Equipment in 1946, the year the company was purchased by Deere. *Deere Archives*

XMC-04, still on wheels, awaits conversion by Lindeman. *Deere Archives*

A production Model M fitted with an experimental adjustable front axle in July 1947. *Deere Archives*

An experimental industrial Model M, the Model MI, in August 1947. *Deere Archives*

Two high-clearance tractors: the 40V, with 26in clearance, and the 40H, with 32in clearance, evolved from the 40XSS and 40XH prototypes. Utility and wide-low versions of the Model 40 were also built.

Within months following the production of the first Model 50 and 60 tractors, Deere's management committed to the development of an entirely new line of tractors, based on a revolutionary series of four- and six-cylinder engines (revolutionary for Deere, at least). Their gestation, detailed in the next section, would take until the end of the decade.

In the meantime, engineers at Waterloo and Dubuque also managed the development of the 20 and 30 Series tractors. Basically updated versions of the Models 40, 50, 60, 70, and 80 (the Model R's replacement), these tractors represented a transition to bigger and better things.

Beyond Two Cylinders: New Generation of Power Prototypes: Models OX, OY, and OZ

As previously noted, Deere engineers had initially considered four-cylinder engines for both the Model D and Model A. However, in both cases, the prevailing economic circumstances dictated that the company carry on with its existing two-cylinder motor. By 1950, however, economic circumstances had changed. Tractor sales hit record levels in the immediate post-war years, with U.S. production for the period 1945 to 1950 exceeding the total production for the period of 1925 to 1945.

Deere had decidedly benefited from the boom. Sales throughout the company were strong. Yet, while the company made gains in market share for much of its product line, its share of tractor sales had not advanced against that of International Harvester. To many within Deere management, it was clear any

Above and right: XMT-01, the experimental Model MT, photographed in August 1947 with single and double front tricycle wheels. *Deere Archives*

gains against IH could only follow a dramatic change in the Deere lineup of tractors. After all, in 1950, Deere's perennial best sellers, the A and B, were nearly two decades old; the Model D nearly three decades old.

As horsepower has always been the name of the game, it is hardly surprising that the two-cylinder engine once again came up for review. Although Deere had improved upon and gained more power from its engines over the years, there were practical limits to the amount of horsepower the company could wring from two-cylinders.

The fate of the two-cylinder engine was undoubtedly decided at the highest levels of the company, as significant investments in manpower and money were required to develop a new engine. I did not find exact details as to when it was decided to build not just a new engine, but a completely new line of tractors. Nonetheless, what I did discover is fascinating.

By 1950, an in-line four-cylinder tractor mock-up was built at Waterloo. For whatever reasons, it never progressed to the prototype stage. It was April 1953 before work began on the new line of tractors. Code named OY, OX, and OZ, the tractors were designed to carry V-4, V-6, and V-8 engines, respectively.

By January 1955, a gas V-6 engine had been placed in an OX mockup, and within a few months, the first OX prototype was built. It featured a V-6 diesel engine. The choice of V engines made sense, in that a V engine was more compact than an in-line engine of the same displacement. However, problems with the choice of a V engine soon developed. Be-

JOHN DEERE

XMT-03, fitted with an experimental heater, photographed at Laredo in January 1948. *Deere Archives*

A production Model MT, fitted with an experimental three-point hitch, photographed at Midvale Farm, Tucson in January 1952. *Deere Archives*

cause the engine was placed directly ahead of the operator, heat from the tractor's exhaust manifolds, which were fitted to the outside of the engine, proved unbearable for the operator. Further, the width of the engine interfered with the operator's forward vision when cultivating. While the first problem could very likely have been corrected, the second could not. Consequently, the V engines were dropped in favor of, what was then, more conventional in-line four- and six-cylinder designs.

In their book, *John Deere Tractors and Equipment Volume Two 1960-1990* (American Society of Agricultural Engineers, 1991), Don Macmillan and Roy Harrington wrote that the first OX mockup with an in-line six-cylinder engine appeared by July 1956. They explained that the narrowness of the tractor's frame required compromises in the engine's design. The frame was widened in a 1958 prototype, which according to Macmillan and Harrington, "permitted the redesign of the engine into what is known today as the 400 series."

An early MX prototype during 1941 field trials. *Deere Archives*

The first MX prototype of 1938. *Theo Brown's Journals, Worcester Polytechnic Institute*

One of five MX diesels built in 1944. *Deere Archives*

One of the final series of MX prototypes built in 1947. *Deere Archives*

The "New Generation of Power" tractors, as they became known, were significantly more than traditional John Deere tractors equipped with new engines. The company had invested $50 million dollars in their development, and the results were a redefinition of the farm tractor.

Styling was executed by Henry Dreyfuss Associates, with particular attention paid to operator comfort. Everything from the angle of the steering wheel to the optimal position of the operator's legs were addressed. A new hydraulics system incorporated a variable-displacement pump that served the requirements of a hydraulically activated multi-speed transmission, power steering and braking systems, and provided up to three live circuits for equipment control. A dual-speed power take-off system and a single-speed front power take-off system were also offered.

While beloved by many, the last Waterloo-built two-cylinder tractor rolled off the line in July 1960. Although Deere did not immediately overcome the sales lead held by IH tractors, there is no question that the first of Waterloo's New Generation of Power (Models 3010 [OY], 4010 [OX], and 5010 [OZ]) had advanced Deere's standing in the industry.

The Model AX with a tandem disk harrow. April 1950.
Deere Archives

Model 40T prototype photographed in July 1952. *Deere Archives*

Busted! 40XMT wide front prototype. *Deere Archives*

Model 40C prototype No. XMC-12. *Deere Archives*

High-Clearance (26in) 40H prototype, 40XS-04, being unloaded upon arrival in Texas. October 1954. *Deere Archives*

40XH-07, Hi-Crop prototype, with No. 472 plow under test in Laredo, Texas, December 1954. Hoppers on the front hold fertilizer for vegetable cultivator. *Deere Archives*

JOHN DEERE
A

The first prototype on the way to the New Generation of Power. A gasoline four-cylinder mockup of 1950. *Deere Archives*

A V-4 mockup of May 1954 featured planetary final drives. *Deere Archives*

Inline six-cylinder OX prototype. *Deere Archives*

OX-602T, a January 1956 OX prototype, featured a heat shield over the exhaust manifold. *Deere Archives*

April 1955 4010 prototype, OX-600, with V-6 diesel.
Deere Archives

JOHN DEERE
3010

Preproduction 3010 gas tractor in November of 1959.
Deere Archives

Chapter 8

Model 101, Model 102, and Model 202

In an era of 300-plus horsepower diesel tractors and 24-speed powershift transmissions, it is difficult to imagine a time when John Deere was working to develop 14 to 15hp gasoline-powered, three-speed general purpose and utility tractors. Yet in 1940, when 40 percent of American farms were smaller than 50 acres and 90 percent of the general purpose tractors built were rated below 30hp, there was significant potential for such tractors. Deere already built two, the Models LA and H, when Theo Brown began work on a prototype he described as a "full vision, four-wheel, general purpose-type tractor." Originally called the "one-row cultivating tractor," it became the Model 101.

The first prototype of Theo Brown's All-Visibility tractor. A mockup of the Model 101 was built on a Model LA chassis, December 1941. *Deere Archives*

The 101 was designed to provide optimal unobstructed views of the ground directly in front and behind the operator. The concept began as a sketch drawn on November 4, 1941. A full scale model employing the chassis of a Model LA and a wooden mockup of the Model H engine and transmission was soon built. From this exercise, it was determined that the horizontal construction of the Model H engine and transmission posed structural complications which left little room for the transmission.

By January 1941, a functioning prototype was built employing an early Model Y chassis. The 14.5hp Model LA motor was placed as far as possible to the rear, with the radiator positioned behind the motor. The seat was placed over the motor and the steering post was repositioned. The original 61in wheelbase was lengthened to 73in, which moved the cultivator shovels forward. The tractor was fitted with 8x24in rear wheels and a three-speed transmission.

A third prototype of the 101 was built in 1944 employing a 15hp Wisconsin VE-4 air-cooled engine, the same engine Deere used in pick-up hay balers. It was this third prototype that Brown proposed be put into production. On January 30, 1945, a group of Deere executives met in the field to evaluate three prototypes, each of which was being considered for production: the 101, the diesel MX, and the 69. The 69 and 101 were demonstrated against Farmall A and Allis-Chalmers B tractors, while the MX was evaluated without such comparisons. The group, which included Charles Wiman, Frank Silloway, Elmer McCormick, and Theo Brown, concluded that all three tractors should be added to the Deere line.

The 101 was never placed into production, although Brown, with the support of Charles Wiman, continued development of what he

Above and left: The second 101 prototype was built on an early Model Y chassis with its wheelbase extended. *Deere Archives*

termed the "All-Visibility Tractor" concept. By 1950, he had developed the models 102 and 202, two-row versions of the 101. The 102 featured a tricycle front end; the 202 a wide front, but like their predecessor, neither tractor saw production.

Theo Brown's "Low Down" Tractor

Operator safety has always been of concern to farm equipment manufacturers, as well as to the American Society of Agricultural Engineers (ASAE). For over 70 years, they have worked together to promote industry-wide safety standards.

Throughout his career, Theo Brown represented Deere & Company on a variety of safety committees. Beginning in the late 1930s, he committed much of his time to designing safety features, which included a "dead man" engine cutoff, and a safety cab and platform. The issue of tractor stability was of particular concern to Brown, and in 1947 he designed a low center of gravity tractor he called the "Low Down" Tricycle Tractor.

Brown's objective was to build a one or two-plow general purpose tractor that would provide optimum forward vision for cultivating and an extremely stable platform. While Brown considered modifying the existing Model B, neither its engine nor transmission was adaptable to the final design. The Low Down prototype was, in fact, a hybrid of the Fordson 2N and Deere Model H: a Fordson engine and transmission, with the transmission case shortened by 15 inches; and 24in diameter Model H wheels, later changed to 28in wheels. The tractor featured a 70in wheelbase and 14in ground clearance.

As with Brown's models 101, 102, and 202, the Low Down tractor was never put into production. Brown's final report on development of the Low Down Tricycle Tractor is on file at the Deere & Company Archives. Only one prototype was built and sent into the field for trials. I found no record of its final disposition.

Theo Brown's Fertile Mind

Theo Brown was a brilliant man dedicated to Deere & Company and to the betterment of the machinery it manufactured. His journals are replete with notes and drawings that attest to his prolific genius (at his death, he held more than 100 U.S. patents). As with any engineer, not all of Brown's ideas—even some that were patented—found their way into use. He was, nonetheless, compelled to record them. Its our good fortune that he did, his journals offer a marvelous insight into the workings of his mind.

Four-Wheel Steering

On November 27, 1929, at the time Brown was working to develop the first hydraulic power lift, he recorded the following:

The 101 ably handled a twelve-inch plow. *Deere Archives*

The Model 101 with its Dreyfuss-designed sheet metal.
Deere Archives

"Thought of this idea today. It is in substance to equip a tractor with a number of hydraulic cylinders to be fed by one pump **A** and with controls **I** and pivotal tractor wheels to:

1—lift implement by hydraulic power through cylinder **F** and control **O.**

2—steer front wheels by cylinder **D** and control **S.**

3—move cultivator **I** through links **J** at same time as steering.

4—steer front and rear wheels through **D** and **E** and controls **P** and **S.** These controls can be interconnected as shown or can work independently.

5—cultivator **K** can be shifted laterally by control **N** through cylinder **G.** In this case, link **M** would be disconnected.

6—wheels **B** can be steered through control **P** and cylinder **E,** and (cultivator) **K** by disconnecting links **M** and **T** (which) can be locked to tractor frame. When moved laterally links **L** pivot.

7—wheels **B** can steer and through link **M** cultivator **K** can move laterally with steering of wheels."

Variable Front Wheel Tread

During the fall of 1930, Brown was working to develop a means of varying the rear tread width of the GP. On September 15 he wrote:

"Thought of this idea today. It is on a tractor to have a front axle hinged in center so that the tread of wheels can be varied. In other words, to make a wide tread or narrow tread tractor with one front axle and having means on rear axle to vary tread there."

Model B Cab

Between May and November 1939, Brown drew several proposals for cabs for the Model B. His drawings of October 4, 1939, were accompanied by the simple statement that "still, cabs are noisy."

Above and right: The third and final 101 prototype featured a Wisconsin air-cooled four-cylinder engine. April and December 1944. *Deere Archives*

The "all-visibility" concept executed as a two-row tricycle tractor, the Model 102. *Deere Archives*

Left: The Model 101 was designed primarily as a one-row cultivating tractor suited to the needs of the small farmer. *Deere Archives*

The Model 202 photographed behind a Ford 8N. *Deere Archives*

The Model 202 featured a wide front axle. *Deere Archives*

Theo Brown's sketch of "what might be done" with a low center of gravity tractor. *Deere Archives*

Brown's Low Down Tractor of 1947 featured a 70in wheelbase and 14in ground clearance. *Deere Archives*

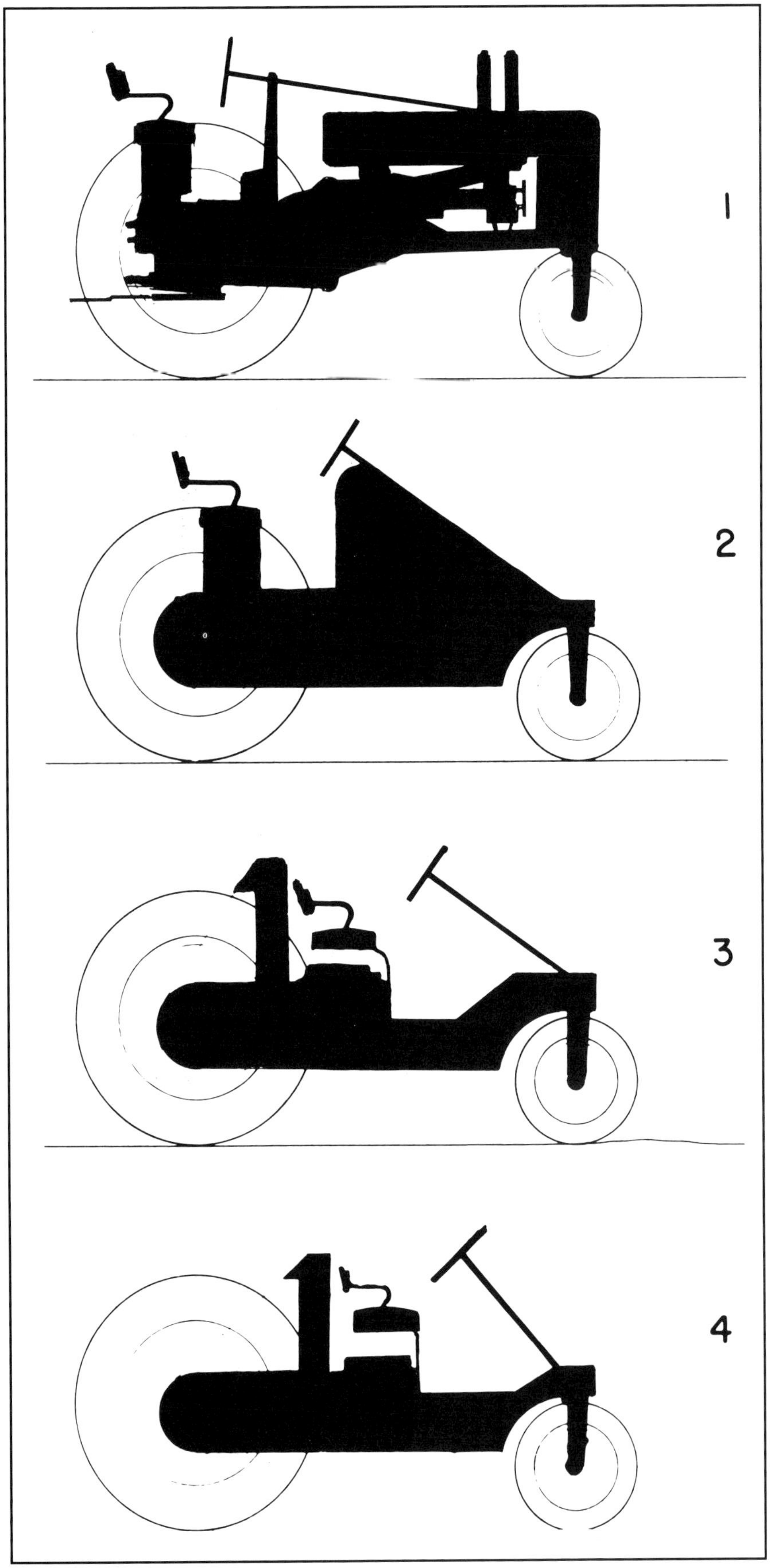

Left and right: Theo Brown created silhouettes to illustrate the possibilities of modifying a Deere Model B (No. 1) and a Fordson 2N (No. 8) to create low center of gravity tractors. No. 5 and No. 6 are based on the Deere B, No. 7 is based on the Fordson, and No. 2, No. 3, and No. 4 are other possibilities. *Deere Archives*

5

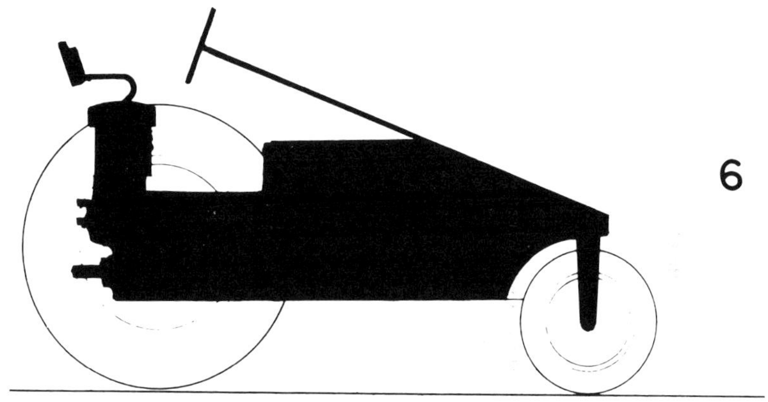

6

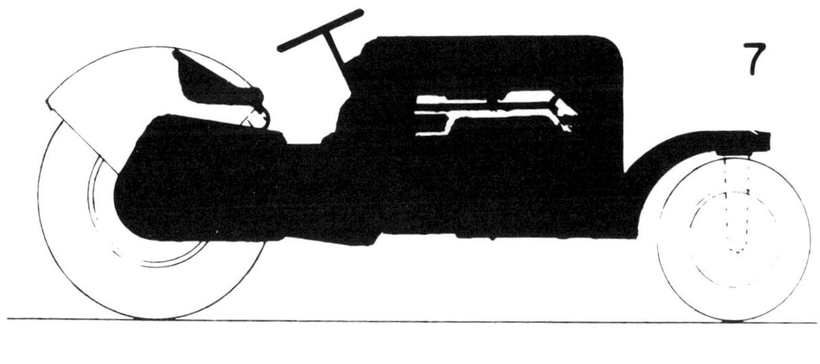

7

8

Above and left: Two views of the Low Down Tractor demonstrate the operator's unobstructed forward view when cultivating. *Deere Archives*

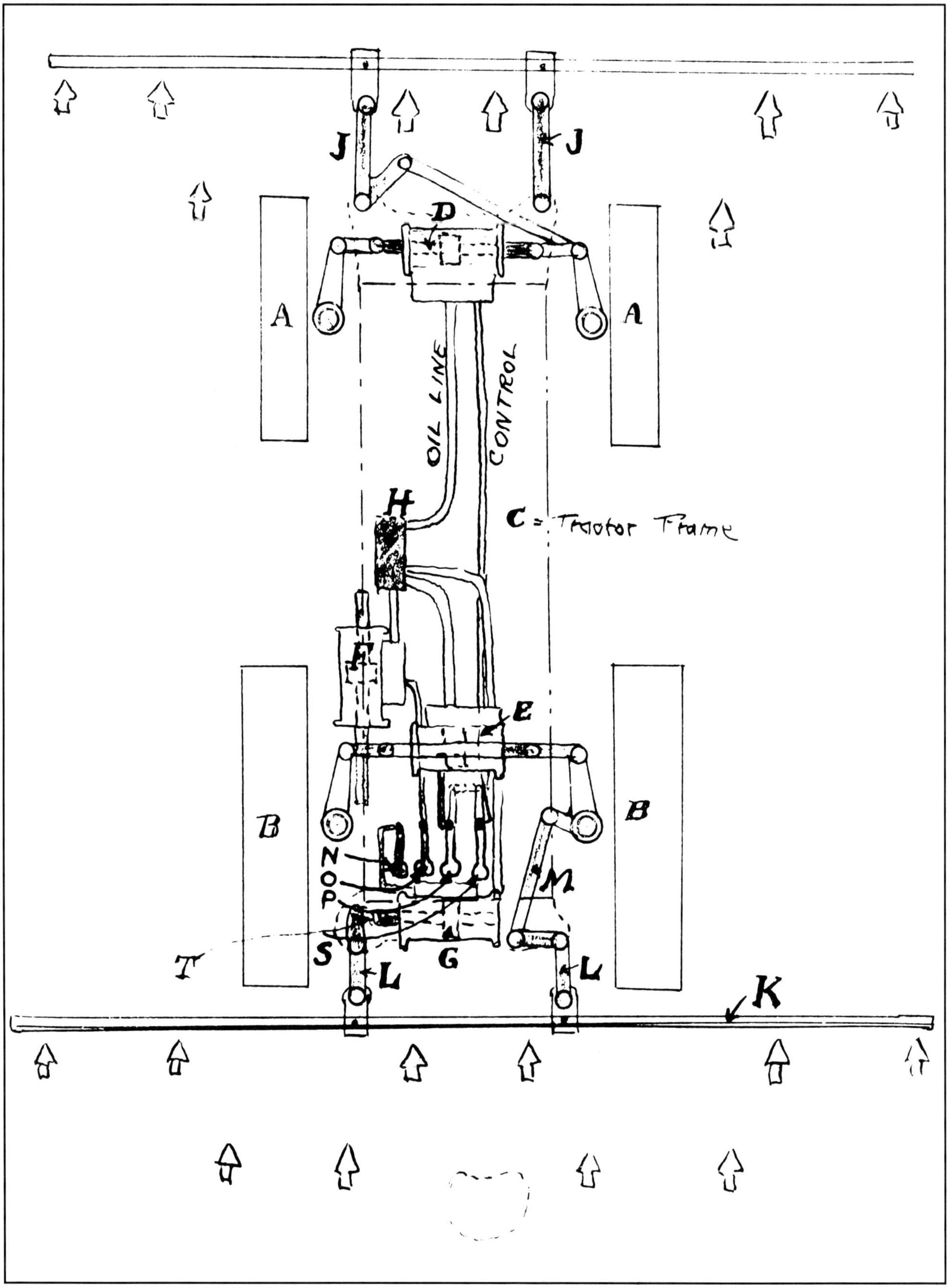

Four-wheel steering. See page 116 for description. *Deere Archives*

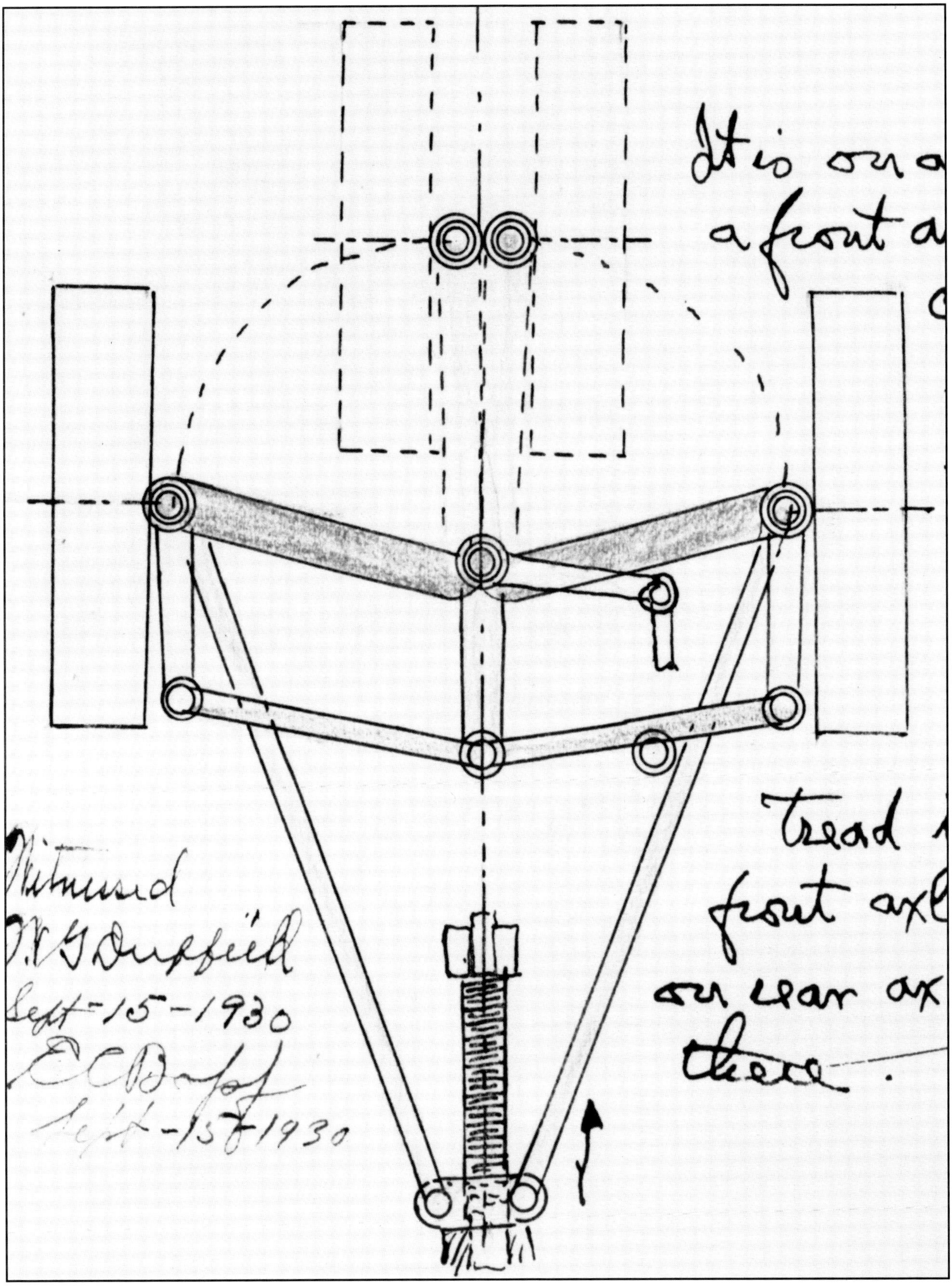

Variable front wheel tread. See page 117 for description. *Deere Archives*

Above, below, and opposite: Model B Cab. See page 117 for description. *Deere Archives*

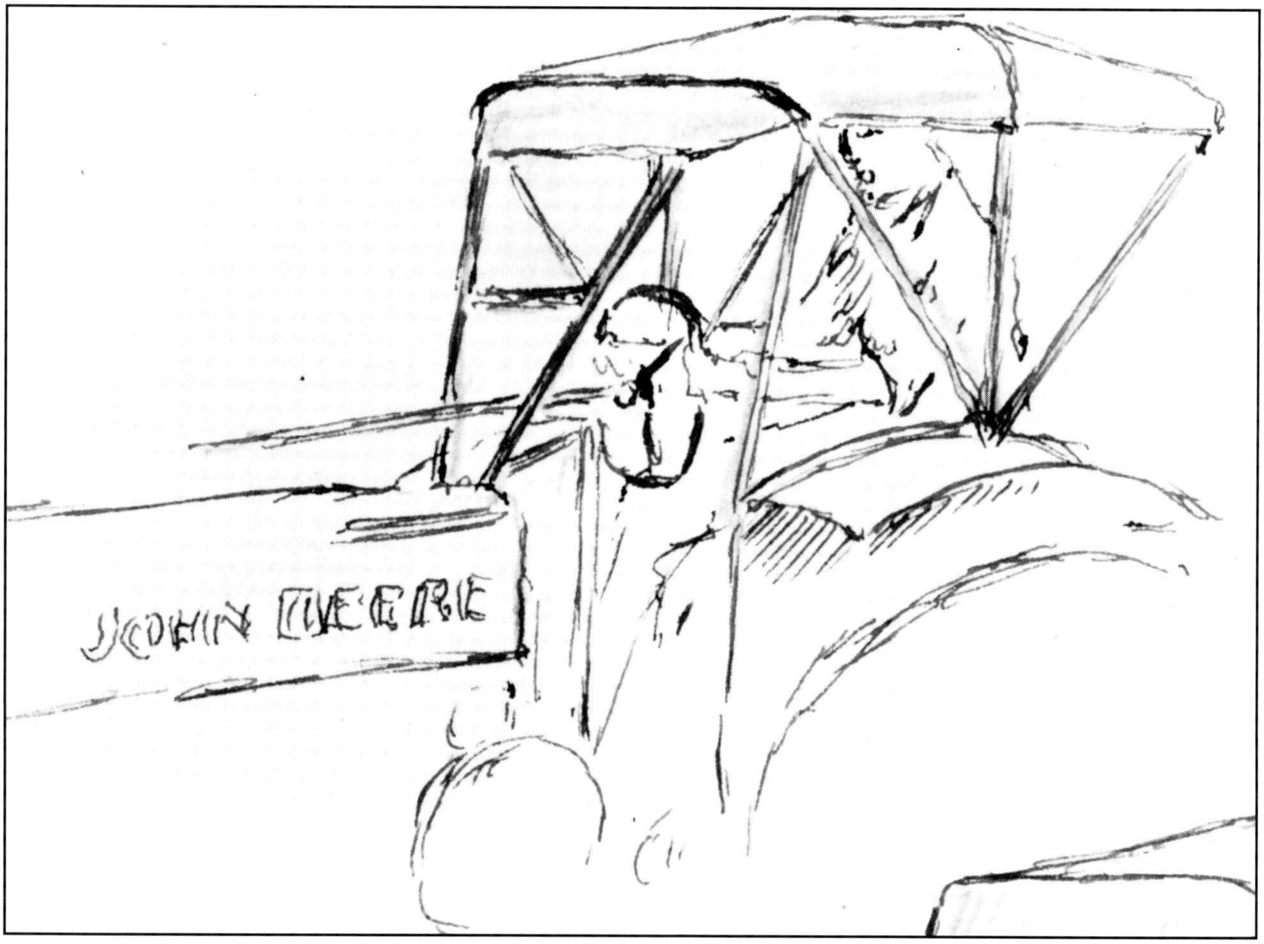

Deere Archives

Deere Archives

Chapter 9

The Henry Dreyfuss Style

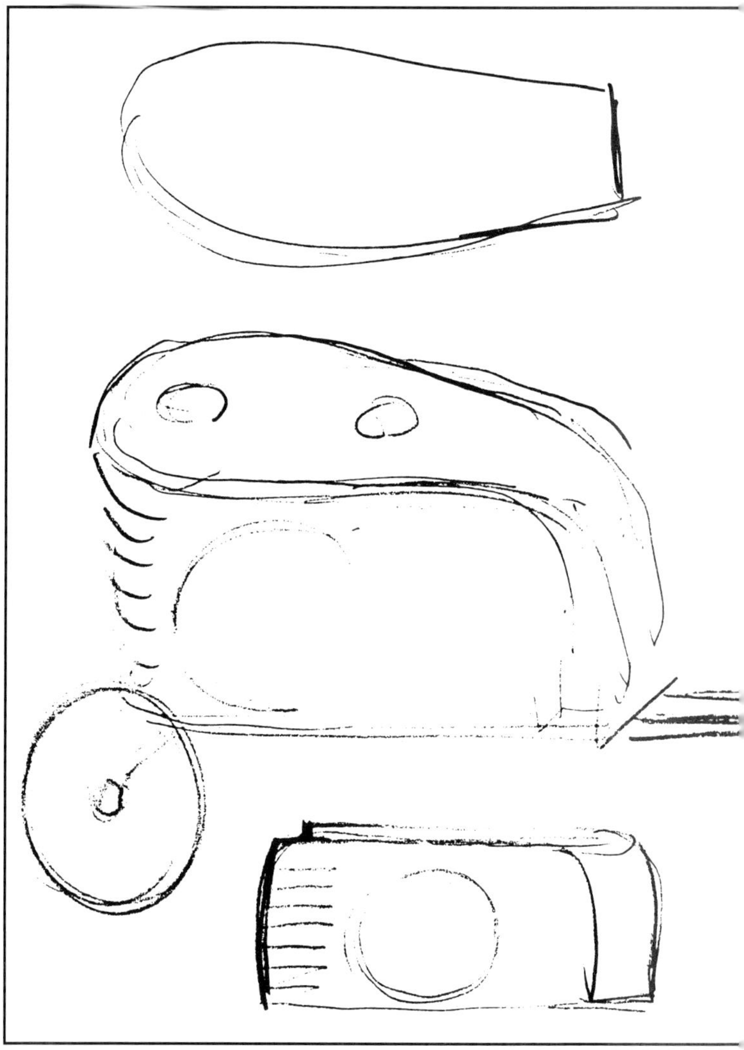

Henry Dreyfuss, the industrial designer, was first hired by Deere & Company in August 1937 to redesign Deere's tractors. His subsequent designs ushered in the era of the "styled" John Deere tractor.

To Dreyfuss, design was not merely a function of appearance. He sought to balance a product's allure with its utility, safety, and cost of production. The Dreyfuss philosophy was embraced by Deere, and for nearly four decades Dreyfuss Associates were responsible for the distinctive look of John Deere tractors.

The Deere Archives have preserved a number of sketches, drawings, and photographs of Dreyfuss's designs dating from the 1930s to the 1970s. I have selected a few for reproduction, as they are intriguing examples of his work.

Model X-61

The process of designing a tractor involves many steps, ranging from a concept captured in a pencil sketch to full-blown working prototypes. At various stages, designers frequently make use of scale models. In 1970, the designers at Dreyfuss Associates created the X-61, a 1/4 scale mockup. The X-61 cab offered the operator unobstructed views in almost all directions through extensive use of glass. Entrance to the cab was gained through sliding doors in the front. Unfortunately, its most unique feature was neither illustrated in the photograph nor was it documented in writing. According to the lean note accompanying the photo, the rear wheels of the X-61 "swing out on large hinges to increase tread width, [which] also shortens the wheelbase."

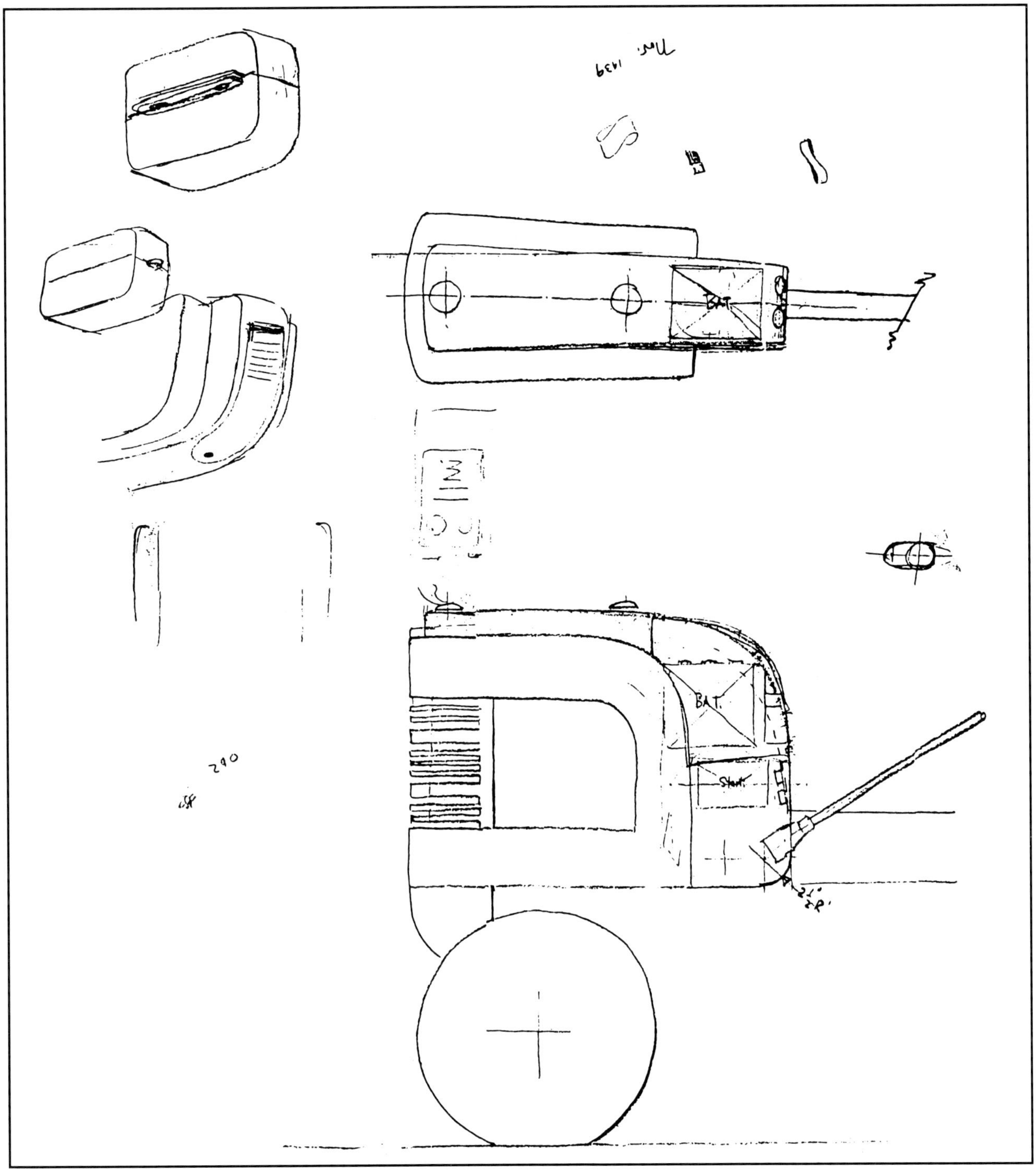

Above and left: Approximately 1,500 units of the Model L were built along the lines of the "unstyled" Model 62, before the tractor was restyled. These drawings by Henry Dreyfuss reveal his earliest idea for streamlining the L, and his notion of a battery box for the restyled tractor. *Deere Archives*

JOHN DEER

The Model 69 was the prototype for the Model M. This Dreyfuss drawing of 1945 proposed styling which, with minor changes, was adopted in 1947. *Deere Archives*

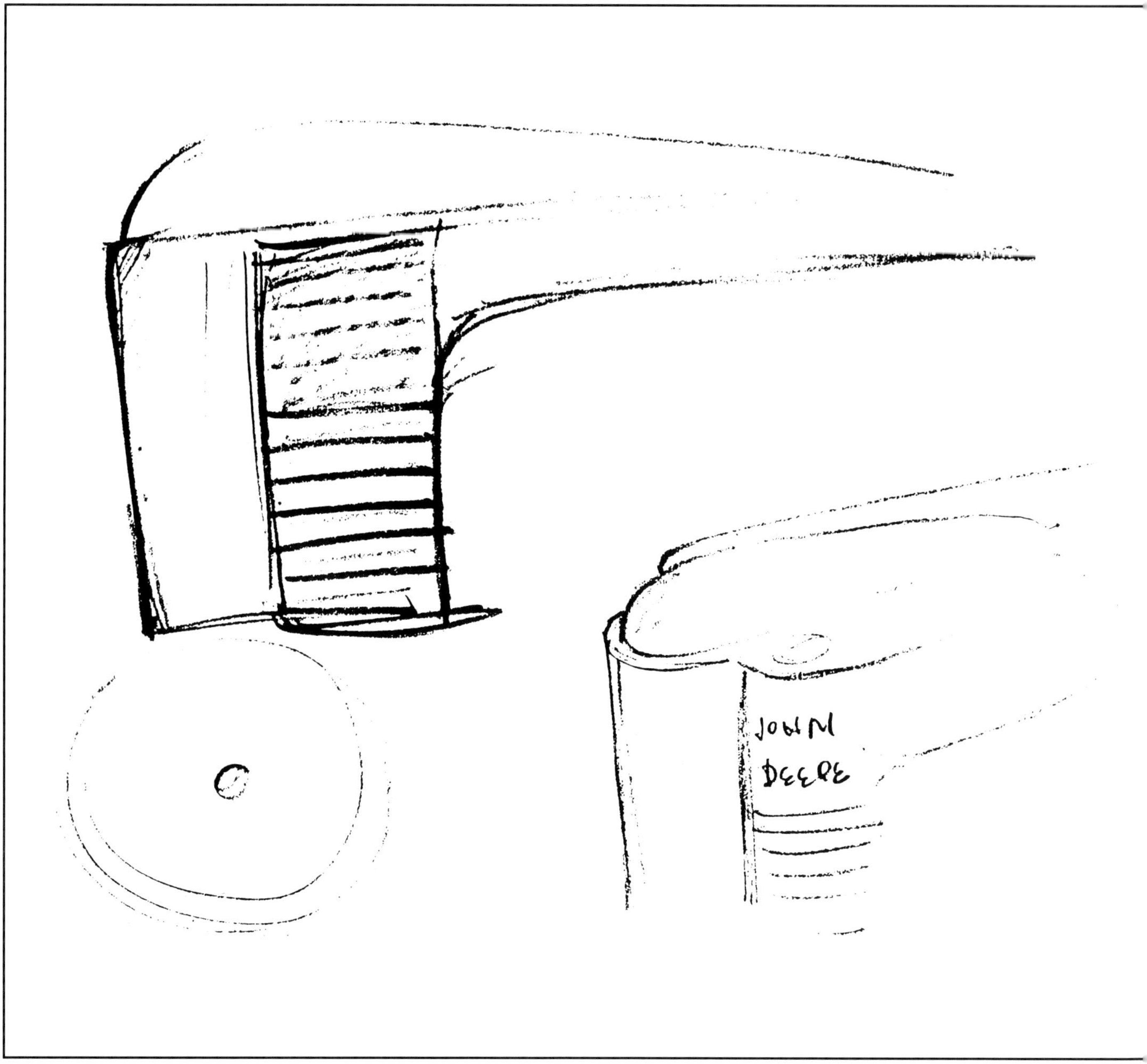

A sketch of two hood lines. Date unknown. *Deere Archives*

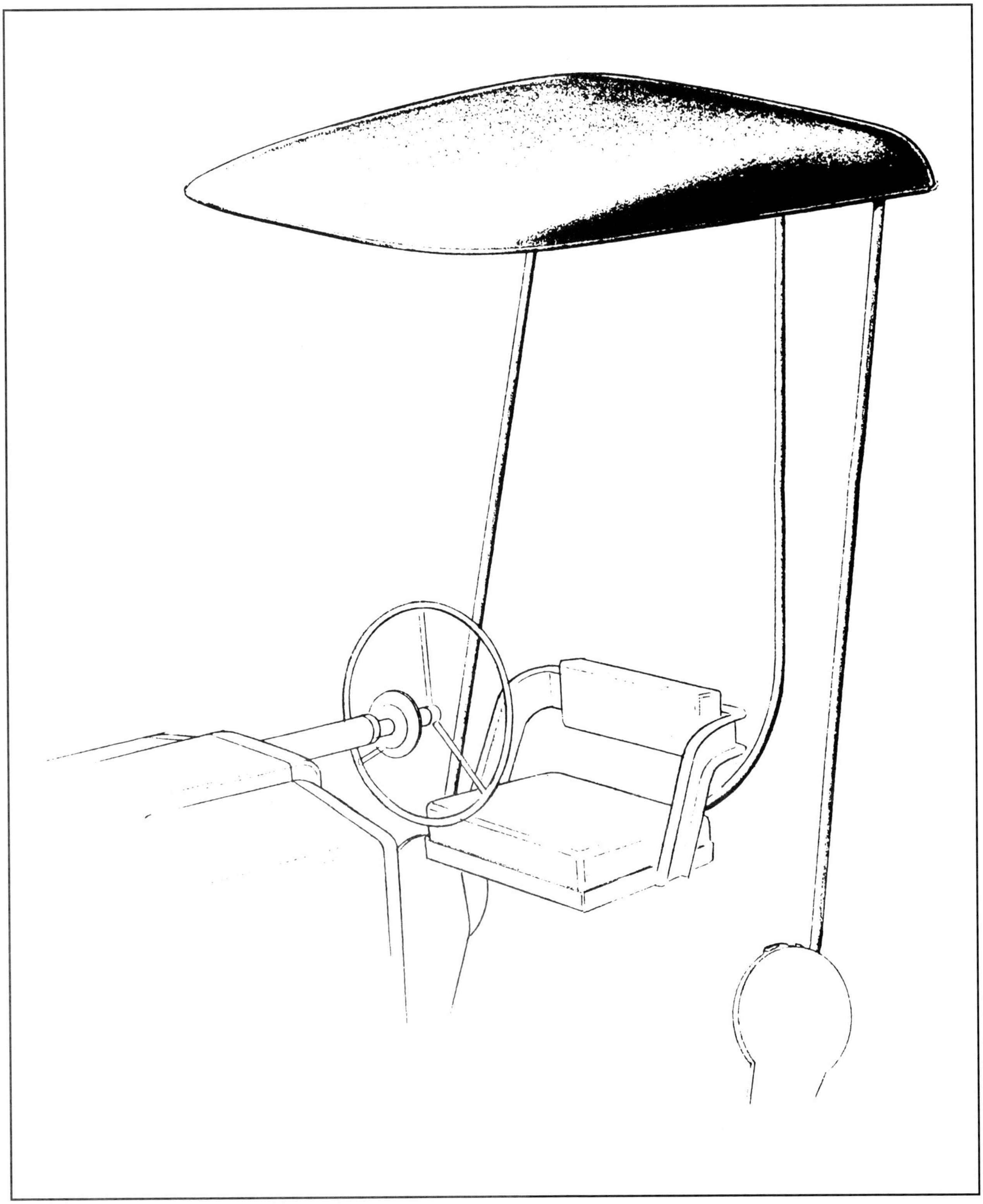

A Dreyfuss-designed canopy for the Model M. *Deere Archives*

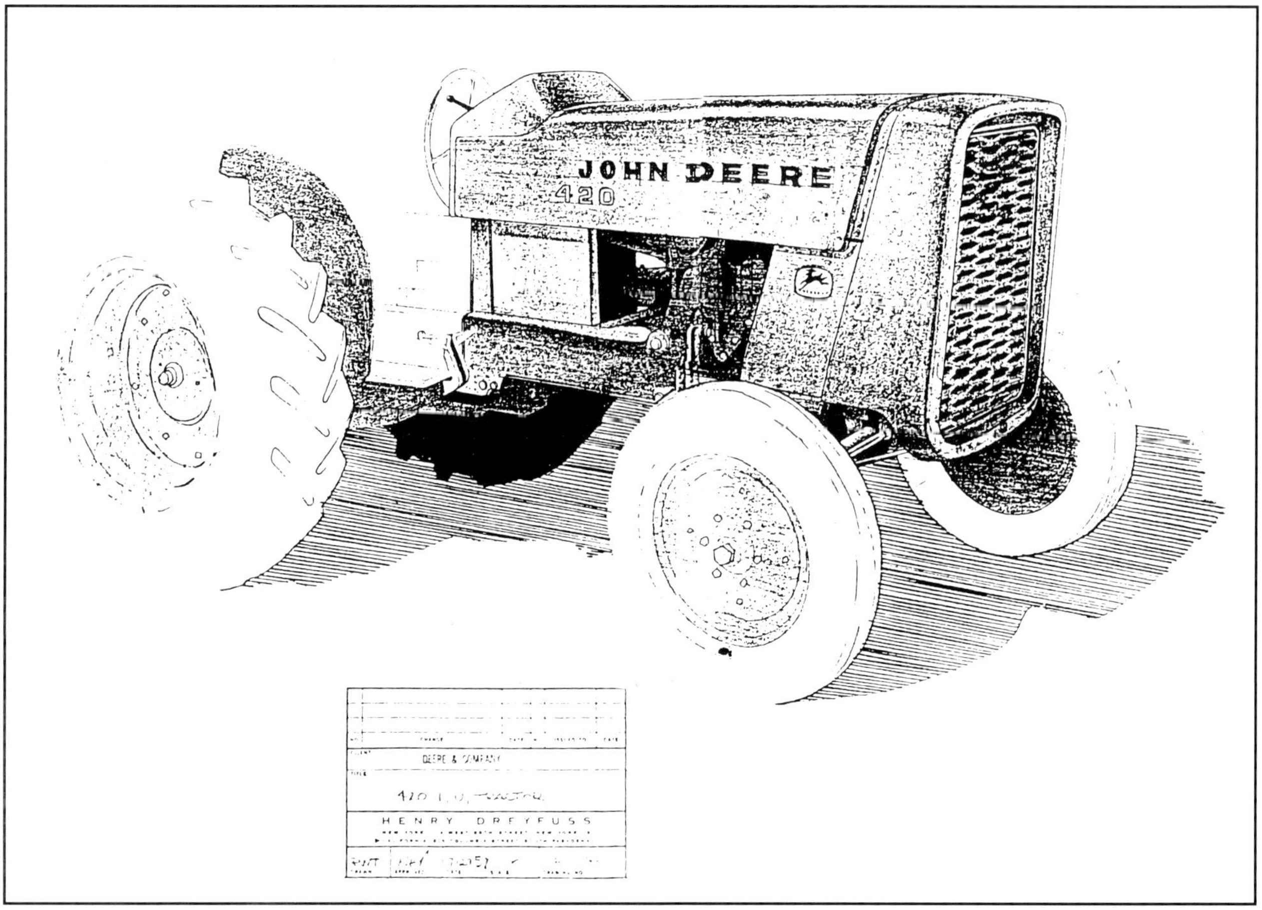

A drawing for the proposed Model 420 IU (Industrial, Utility) tractor of July 20, 1957. *Deere Archives*

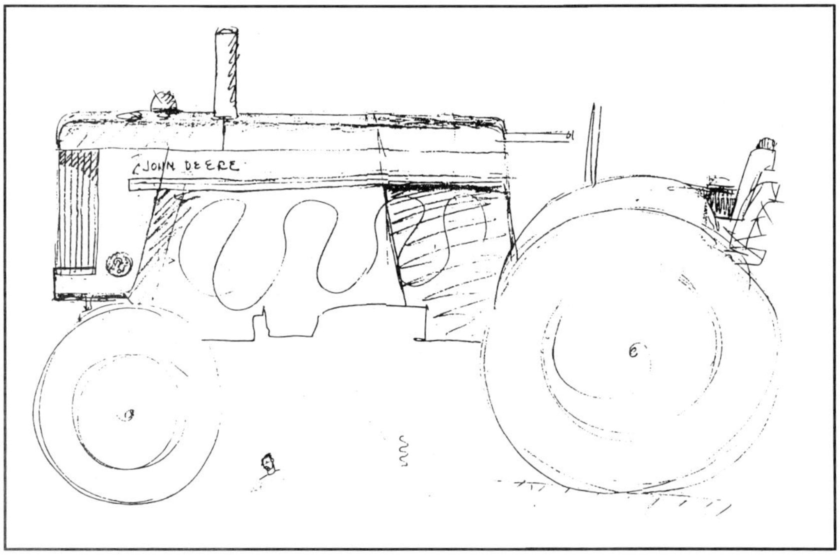

A Dreyfuss sketch dated July 8, 1948, reflects the general changes in styling that would follow introduction of the Model R. *Deere Archives*

Dreyfuss styling applied to the John Deere Lanz prototypes of the 18-PS 100, 28-PS 300, and 36-PS 500 tractors. *Deere Archives*

A 1/4 scale mockup of the model X-61 prototype. *Deere Archives*

Index